# WORKSHOP

## The Complete Beginner's Guide to AI

YOUR A-Z GUIDE TO MASTERING ARTIFICIAL INTELLIGENCE FOR LIFE, WORK, AND BUSINESS

**NO CODING REQUIRED**

---

ONLINE AI CLASS

HANDS-ON EXERCISES

100+ PROMPTS

---

**MILO FOSTER**

Published 2025 | FIRST EDITION

ISBN # 978-1-967210-00-8 eBook

ISBN # 978-1-967210-02-2 paperback

ISBN # 978-1-967210-03-9 hardcover

ISBN # 978-1-967210-04-6 audiobook

For permission requests or business inquiries, please contact: milofoster059@gmail.com

Disclaimer

This book provides general information about artificial intelligence. The author and publisher present this material as-is, without warranties of any kind. While reasonable efforts were made to ensure accuracy at the time of publication, technology changes quickly. Tools, links, and resources mentioned in this book may evolve or become unavailable.

This book does not provide legal, financial, or professional advice. Before applying any information from this book, consult qualified professionals for your specific needs. The author and publisher are not responsible for any outcomes resulting from the use or misuse of the content.

The author has no control over third-party websites or services mentioned and does not guarantee their accuracy, reliability, or practices. Mention of any product, service, or brand does not imply endorsement.

By reading this book, you agree to use its content at your own risk and acknowledge these limitations.

# Acknowledgments

THIS BOOK WOULD NOT HAVE BEEN POSSIBLE without the insights and experiences shared by professionals across various industries. I am deeply grateful to the clients, business leaders, entrepreneurs, educators, and AI experts who generously contributed their perspectives.

To protect privacy and confidential business information, some of the individual and company names have been omitted. Instead, you will find references such as "a CEO at a financial firm" or "a director at a marketing agency." These examples reflect real-world experiences and conversations, with minor details adjusted for clarity and anonymity.

I am especially thankful to everyone who took the time to share their knowledge. Your contributions helped bridge the gap between AI theory and its practical applications, making this book a more valuable resource for readers.

"AI shouldn’t be
an exclusive tool
for tech experts.
It should be
accessible to
everyone.
Including you."

Milo Foster

# Table Of CONTENTS

## Before You Get Started

Before you dive in, I want you to know that this book comes with two complimentary bonus tools to help you get even more out of your AI journey.

These extras are included in your book purchase and ready for you to use. Just visit www.funtacularbooks.com and join the mailing list to unlock instant access.

Here's what you'll get:

**100+ AI Prompts: AI Prompt Library for Smarter Interactions** created by Milo Foster

This handy download gives you ready-to-use prompts for tools like ChatGPT, Claude, and Gemini. No tech skills required—just copy, paste, and go.

You'll find helpful prompts for:

- Writing emails, bios, articles, and social posts
- Brainstorming when you're stuck
- Analyzing data and making better decisions
- Planning meetings, projects, or marketing
- Creating lesson plans, customer journeys, and more

You'll also get a simple framework for writing your own prompts, plus tips for fixing AI responses that don't quite hit the mark. It's perfect for beginners and helpful even if you already use AI tools.

No cost. No fluff. We never sell your information or spam your account. **Just practical tools to help you feel more confident and capable with AI.**

**Unlock 100+ Easy AI Prompts for Everyday Use!**

Visit www.funtacularbooks.com/aiprompts

Or scan QR code to download TODAY!

## Online Mini-Course: Future-Proof Your Career with AI

This short, focused course shows you how to start using AI in your daily work—no coding or tech background needed.

It includes:

- 4 quick video lessons that show you what to do and why it works
- Real-world examples from different careers and industries
- Simple activities so you can apply what you learn right away
- A printable worksheet to help you plan your next steps with AI

You can finish the course in under 30 minutes and start using what you learn the same day.

**FUTURE PROOFING YOUR CAREER WITH AI Online Mini Course with LIFETIME ACCESS**

Visit https://bit.ly/AIWORKSHOPCAREER

Or scan QR code to access TODAY!

"We are called to be architects of the future, not its victims."

R. Buckminster Fuller

# How This All Started

IT WAS THE SUMMER OF 1991. I was fifteen, sitting in a movie theater with my dad. I gripped the armrests as I watched the T-1000 melt into liquid metal and transform into a police officer. It was my first time seeing an R-rated movie in a theater: Terminator 2: Judgment Day. It was also my first introduction to artificial intelligence. For two hours, I watched a future where smart machines decided humans were a problem and came up with a simple solution: wipe us out.

I walked out of the theater feeling both excited and uneasy. The idea of AI turning against us, with Skynet becoming self-aware, killer robots on the loose, and humanity on the brink of extinction, stuck with me for years. I imagine some of you grew up with the same sci-fi stories that still shape how you think about AI.

But my experience with technology led me down a different path. After spending nearly 25 years working in information technology, my view of AI completely changed. What I discovered was not the conscious, menacing machines from science fiction. Instead, I saw powerful tools created by people to solve specific problems.

In reality, AI today looks nothing like the Terminator. There are no sentient machines plotting to destroy us. Instead, I have seen AI help doctors catch diseases earlier, assist people with disabilities in navigating the world, and give small businesses the ability to compete with larger companies.

What Hollywood shows as mystical, godlike intelligence is actually just advanced pattern recognition. It is incredibly complex math that spots relationships in data that humans might miss. It is not magic, and it is definitely not alive. Even the most advanced AI systems today have no awareness, no desires, and no ability to turn against their programming. They are simply tools. Impressive ones, but still just tools made to enhance human abilities, not replace them.

This difference is important. When people misunderstand AI, it creates unnecessary fear and leads to unrealistic expectations. But when we see AI for what it really is, advanced software rather than electronic life, we can approach it with the right mix of caution and confidence. We can appreciate its massive potential while also recognizing its limits.

Throughout my career, I have noticed that the most effective AI applications are not trying to mimic human thought. Instead, they handle specific tasks that support what humans do best. AI can scan millions of medical images without tiring, but it still takes a human doctor to combine those results with a patient's history, preferences, and unique circumstances.

This partnership between human and artificial intelligence, rather than competition or replacement, is the proper story of AI. It is also a story Hollywood rarely tells. And it is the story I hope to share with you throughout this book. By learning what AI actually is, how it works, and how people are already using it, you will move past both fear and hype. You will gain a clear, practical understanding of this powerful technology.

In the chapters ahead, we will explore AI not as science fiction, but as a set of practical tools that are already changing how we work, learn, and solve problems. You will see that you do not need a technical background to use AI today.

My name is Milo Foster, and I have worked in information technology for almost 25 years. I have done everything from fixing servers in the middle of the night to setting up computer systems for large companies with thousands of employees. Most of my work has been as an IT consultant for small businesses, local government offices, and healthcare organizations—places that need technology to solve problems but don't have big budgets or dedicated tech teams.

What interests me most is how different organizations adopt new technology in different ways. A small factory might quickly start using automation, while a government office could take years to approve the same tools. A healthcare company might use AI to schedule patient visits but be much more cautious about using it for medical decisions. I have seen these choices play out firsthand, along with the successes, failures, and everything in between.

Throughout my career, I have often acted as a translator between technical teams and the people who use the systems. That's exactly what I am doing with this book, explaining AI in a way that makes sense, even if you are not a programmer or data scientist.

For example, I remember sitting with the owner of a small dental practice as she tried to understand how AI could help her business. Her eyes glazed over as the vendor tossed around terms like "neural networks" and "supervised learning algorithms." After the meeting, she turned to me and said,

*"Milo, I just want to know if this will help my patients and save my staff time. Why can't they explain it that way?"*

She was right. That conversation was one of many that inspired me to write this book, to explain AI the way most people actually need it explained.

## Why This Book? Why Now?

AI isn't just another passing technology trend. It's transforming industries, creating new jobs while changing others, and reshaping how we interact with the digital world. Yet while companies race to implement AI and headlines trumpet its advances (or warn of its dangers), it leaves people wondering:

*"What does this mean for me? How can I actually use this stuff?"*

Wait, I'm getting ahead of myself. Let's back up.

The real problem is not just understanding AI. It is about connecting big ideas to useful, everyday uses. It is about changing from thinking "that sounds interesting" to thinking "here is how I can use this right now."

Most AI resources fit into two categories: very technical manuals for programmers or basic overviews that don't help you do anything practical. This book is different. It aims to:

- Explain AI concepts in plain, jargon-free language
- Provide hands-on, no-code tools and projects anyone can use immediately
- Show real-world applications across various fields and interests
- Build your confidence in discussing, using, and implementing AI solutions

Back in 2007, I worked with an early recommendation system. It was simple compared to what we have today, just a basic program that suggested products based on what people bought before. One day, I saw it was making strange suggestions that didn't match well. After looking into the problem for several days, I found that the issue wasn't in the computer code. The problem was how the system understood connections between products that didn't seem related.

Customers buying gardening tools were also often buying a specific brand of chocolate. The store displayed the chocolate near the checkout to encourage impulse buys, not because it related to gardening. The AI spotted the pattern but did not understand the reason behind it, something a person would have noticed right away.

That lesson has stayed with me: even the smartest computer programs do not have the natural understanding that humans have. Today's AI is powerful, but we get the most from it when we know both what it can do and what it cannot do.

## What to Expect (and What Not to Expect)

This is not a textbook about machine learning algorithms. It is not a guide for programming. And it is definitely not a philosophical paper about how AI will change humans in the future (though we will talk a little about these ideas).

Instead, think of this as your practical field guide to the AI landscape, written by someone who's been exploring it for years but remembers what it's like to be a beginner.

In these pages, you'll find:

- Straightforward explanations of how AI actually works
- Step-by-step guides to using no-code AI tools for personal and professional tasks
- Actual stories of AI successes and failures (because we learn from both)
- Practical projects you can complete regardless of your technical background
- Honest assessments of AI's capabilities and limitations

I've structured this book to build your knowledge gradually. We'll start with the fundamentals, explore how AI works behind the scenes, and then dive into practical applications. Along the way, I'll share insights from my years in the tech industry. The kind of real-world wisdom that rarely makes it into books.

## TRY IT YOURSELF

This is important.

Each chapter has a **"Try It Yourself"** section with simple, hands-on exercises. Reading about AI is helpful, but using it makes a bigger impact. These activities are neither complicated nor technical. They are easy to follow for anyone who knows how to use basic digital tools.

## Who This Book Is For

- The curious professional who wants to stay relevant in an AI-driven workplace.
- A small business owner wondering how to leverage AI without hiring a technical team.
- A student exploring potential career paths in a transforming economy.
- The creative person looking to enhance their projects with AI tools.

In short, it's for anyone who's heard about AI and thought:

*"I wish someone would explain this to me in a way that actually makes sense."*

If you've ever felt intimidated by technology discussions, if you've wondered how to separate AI fact from fiction, or if you're simply curious about how these tools might fit into your life, this book is for you.

I hope that by the time you finish this book, you'll not only understand AI better, but you'll have already started using it in ways that enhance your daily life and work. Because AI shouldn't be an exclusive tool for tech experts. It should be accessible to everyone. Including you.

Let's explore AI together. We'll skip the complicated tech talk. No coding skills needed. We'll focus on practical know-how and give you real ways to use the technology that's changing everything around us.

Turn the page, and let's demystify AI, one concept at a time.

"The question of whether a computer can think is no more interesting than the question of whether a submarine can swim."

Edsger W. Dijkstra

CHAPTER 01

# Understanding AI Fundamentals

REMEMBER THAT FIRST TIME YOU USED GPS navigation? That moment when a computerized voice confidently instructed you to "turn right in 500 feet" and somehow knew exactly where you were going? If you're like me, there was something almost remarkable about it. Not because it seemed magical, but because it showed how effectively technology can solve complex problems.

Let's look at how these systems actually work. At its core, artificial intelligence relies on math and data processing. It uses complex algorithms to find patterns and make predictions. Even though AI uses complicated math, the main ideas are simple enough for anyone to learn, regardless of their background in technology.

In this chapter, we'll break down these core concepts and provide a practical foundation for understanding how AI works, what it can do, and, most importantly, how you can use it to solve real-world problems.

## Understanding AI: Key Concepts Explained

Technology often uses complicated words that can be confusing, especially for artificial intelligence. Let's explain the main ideas in a simple way, even if you have never studied computer science.

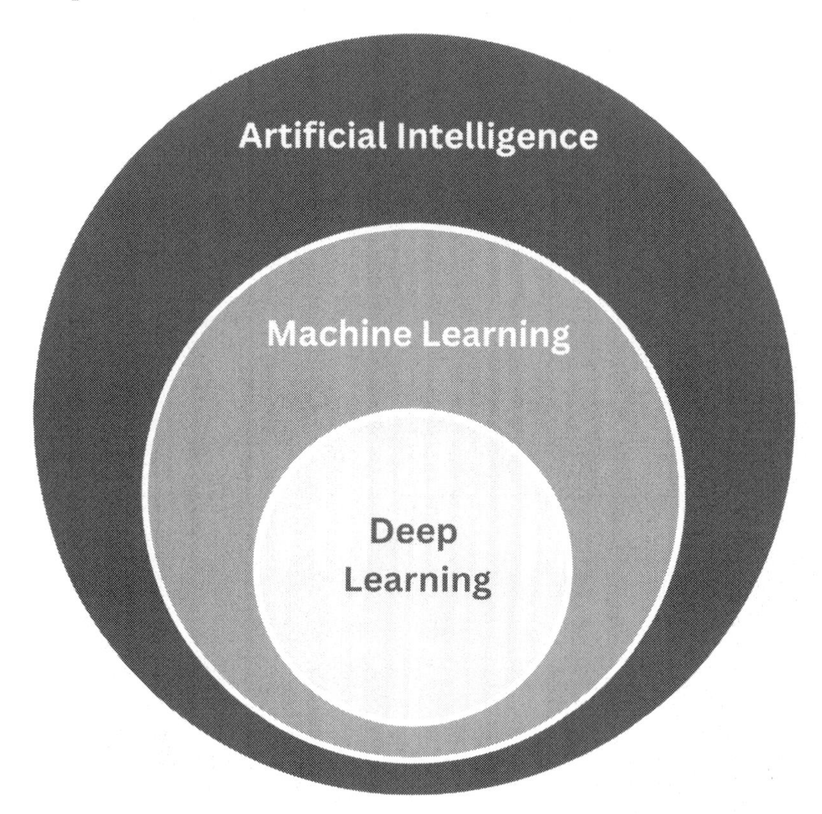


Figure 1.1

**Artificial intelligence (AI)** is the broadest term. It refers to creating computer systems that can do tasks that usually require human intelligence. This doesn't mean machines think like humans, but they can process information, spot patterns, and make decisions in ways that seem smart.

**Machine learning** is a smaller part of AI. If AI is the big idea of making intelligent systems, machine learning is the method that helps computers learn from data without needing to be programmed for every task. It is like teaching a child to recognize different fruits. Instead of listing every detail of an apple, you show them many apples, and they learn to recognize them. Machine learning works the same way by letting computers study large amounts of data to find patterns.

**Deep Learning** is an even more specialized type of machine learning. It uses what we call neural networks, computer systems designed to work a bit like the human brain. These networks have multiple layers that can recognize increasingly complex patterns. The "deep" in deep learning refers to these multiple layers of processing.

An **algorithm** is like a recipe for solving a problem. In AI, algorithms are sets of step-by-step instructions that help computers process information, make decisions, or learn from data. Just as a recipe shows you how to prepare a dish, an algorithm tells a computer how to handle a specific task.

As figure 1.1 shows, these concepts build on each other:

- AI is about creating smart computer systems
- Machine Learning is one way to create AI
- Deep Learning is an advanced type of machine learning
- Algorithms are the specific instructions that make these systems work

The most important thing to remember is that these are tools designed to help humans, not replace them. Each of these technologies is about extending our capabilities, solving complex problems, and creating new possibilities.

## What AI Actually Is (And Isn't)

After exploring the basics of AI, let's look at what makes these systems truly different. Traditional software follows strict, step-by-step rules: when **A** happens, it does **B**. AI works differently. It can recognize patterns, make predictions, and get better over time without needing to be programmed for every situation.

I often explain this with a simple comparison:

> *Traditional software is like giving someone turn-by-turn directions to a place. AI is more like teaching them how to read a map and let them find their own way. It can adapt, learn, and discover new solutions.*

The history of AI is a story of remarkable ambition and gradual progress. When John McCarthy first coined the term in 1955, he and his colleagues believed machines could describe and simulate every aspect of intelligence.1 Pretty ambitious for the 1950s, right?

Early AI was incredibly basic. The first programs could play simple games like checkers or solve basic algebraic problems. These tasks seem trivial now, but they were significant breakthroughs at the time. Cycles of excitement and disappointment have marked the journey of AI, including multiple "AI winters" when funding and interest would dramatically drop.

A pivotal moment came in 2012 with deep learning and neural networks. Suddenly, computers could recognize objects in images almost as accurately as humans. This wasn't just an incremental improvement, it was a fundamental shift in what machines could do.

Today's AI can do impressive things. It can beat top players at complex games, create original content, and solve problems in many areas. But it is important to understand what AI is not. These systems are not alive. They don't have feelings or awareness. As advanced pattern-recognition tools, they simulate understanding but don't grasp concepts like humans.

The most powerful AI systems work alongside human intelligence, not in place of it. They are great at handling large amounts of data, spotting patterns, and making fast calculations. However, they still need human creativity, ethical judgment, and real-world understanding to be truly effective.

## Common Misconceptions About AI

Let's address some practical misconceptions about AI that often lead to confusion when implementing these technologies:

### Misconception #1: AI systems can figure out anything with enough data

Many people wrongly think that AI can solve any problem if it has enough information. This creates false hopes and often leads to disappointment.

In reality, AI is best at solving specific problems, especially those that involve recognizing patterns within clear limits. For example, AI might be excellent at spotting tumors in medical scans or predicting equipment failures. However, it often struggles with tasks that need common sense or a deeper understanding of context.

I once worked with a manufacturing client who expected their new AI system to improve their entire production process automatically. They didn't realize that while the AI was good at scheduling and quality control, it couldn't manage workplace culture, employee preferences, or unexpected problems without being specifically programmed for those issues.

This limitation isn't a failure of AI, but a reflection of its specialized nature. Understanding what problems AI is well-suited to solve, and which ones it isn't, is crucial for successful implementation.

### Misconception #2: AI can replace all jobs

The second big misconception is that AI will take everyone's jobs. This one is a bit more complex because, in the future, AI will definitely change the way we work. But will it completely replace jobs?

That's an oversimplification.

I've helped many organizations use AI systems, and not one aimed to fire workers. Most AI projects focus on making specific tasks easier, not replacing entire jobs.

Think about it this way: when spreadsheet software became common, it didn't replace accountants. It changed their jobs, freeing them from doing calculations by hand and letting them focus on analysis and planning. In the same way, AI helps humans work better rather than taking over completely.

Research from McKinsey suggests by 2030, about 14% of the global workforce may need to switch occupational categories because of automation, including AI.[2] Now that's significant, but it's not the job apocalypse that makes for good headlines.

## Misconception #3: AI is infallible

The third misconception is that AI never makes mistakes or is completely fair. This is especially dangerous because it can make people trust AI systems too much without proper checking.

AI systems learn from data, and that data often has human biases in it. An AI trained using biased hiring decisions will probably continue those same biases. An AI trained on old medical data might give worse advice for groups that weren't well represented in that data.

I worked with a retail chain that wanted to use AI to manage inventory. The system kept suggesting they stock too much of certain products in some stores. When we looked into it, we found the AI had noticed a pattern in past sales—certain products sold well during specific weather conditions. The issue was that this pattern came from one unusual year with extreme weather. The AI saw the connection, but it didn't understand the real reason behind it. It needed human judgment to make sense of the data.

This brings us to an important truth: AI systems aren't objective oracles. These statistical tools analyze and reflect the data they have trained on, for better or worse.

## Key Components of AI Systems

Now that we've cleared up some misconceptions, let's look at what actually makes AI systems work. There are four key components:

1. Data: The Foundation of Learning
2. Algorithms: The Processing Engine
3. Training and Testing Processes
4. User Interfaces

### Data: The Foundation of Learning

Let's start with data, because without it, modern AI simply doesn't exist. The best algorithm in the world can't perform well without good data to learn from.

Imagine trying to learn a language if you only had three sentences as examples. You might memorize those perfectly, but you would struggle in an actual conversation. AI faces the same challenge. It needs sufficient examples to learn meaningful patterns.

It's not just about having more examples. The type and variety of examples matter too. If an AI only learns from certain kinds of examples, it won't know what to do when it sees something new or different. So having examples that show all the different situations the AI might face is really important.

I worked with a factory that wanted to use AI to improve how it checked product quality on the assembly line. They expected the system to quickly spot flaws and reduce the need for manual inspection. However, the first round of results fell short of expectations. The AI struggled with anything subtle and missed small issues that human workers would normally catch without much effort.

After digging into the problem, we realized the training data only included examples of perfect products and ones with major, obvious damage. The AI had no idea what to do with anything in between. It had never seen minor scratches, slight dents, or other real-world imperfections, so it simply ignored them. Once we added a wider range of sample images showing those smaller flaws, the AI finally started to improve. It became more accurate and useful for everyday quality checks.

ETHICAL SPOTLIGHT

**Data Ownership and Consent**

As we look at how AI systems learn from data, an important question comes up: who owns the data used to train these systems? When you use AI-powered services, your interactions often become part of the training data that improves future versions. Companies have a responsibility to be clear about how they use this information and to get proper consent. As an AI user, it is worth thinking about what happens to your data after the interaction ends.

## Algorithms: The Processing Engine

Next, we have algorithms. These are the step-by-step instructions that help AI learn from information and make predictions. Think of algorithms as the recipes that tell the AI how to process data and come up with answers.

Algorithms come in many types, from basic decision trees to complicated neural networks. Each type works better for some problems than others.

For instance, some algorithms are great at classification (is this email spam?), while others excel at regression problems (what will this house sell for?), and still others at reinforcement learning (what's the best move in this game?).

### Training and Testing: The Learning Process

The third component is the training and testing process. This is where data meets algorithms to create functional AI systems. During training, the AI looks at examples and changes itself to better find patterns. It's like when you practice something new - the AI gets better at its job the more it practices.

Testing is equally important. It's where we evaluate how well the AI performs on data it hasn't seen before. This helps determine if the system has truly learned useful patterns or has just memorized the training examples (a problem called "overfitting").

### User Interface: The Human Connection

Finally, we have user interfaces, which connect AI systems with the people using them. Users can ignore even the smartest AI if the interface is clunky. On the other hand, a simpler AI with an easy-to-use design can become essential.

Take Siri or Alexa, for example. These assistants let you talk to them just like you'd talk to a person. You don't need to be a tech expert to use them. Anyone who can speak can give them commands. That's what makes them so amazing.

## How AI Actually Works

When we talk about artificial intelligence, neural networks are like the system's brain. Just as human brains learn by linking unique pieces of information, computer neural networks do something similar. You can imagine them as a team of problem-solvers working together. Each one passes information to the next, constantly adjusting and improving along the way.

Neural networks start with raw data entering through the input layer. This is like the first group of workers receiving the initial information. As the data moves through more layers, called hidden layers, something interesting happens. These layers contain interconnected nodes, which act like mini processing units. Each node takes in information from the previous layer, performs a calculation, and sends the result to the next layer.

The first hidden layer might recognize simple patterns like edges or basic shapes. As the information flows through more hidden layers, the network can identify increasingly complex concepts by combining the patterns from previous layers. This is because each hidden layer builds upon the knowledge of the layers before it.

During training, the neural network adjusts the strength of the connections between nodes based on how well it classifies the input data. If the network makes a mistake, it updates these connections to improve its accuracy. This process of adjusting internal connections is how the neural network learns and gets better at its task.

It is like how a child first learns to recognize basic shapes before understanding what a full dog or cat looks like. As the child's brain builds on simpler ideas, they develop a more advanced understanding. In the same way, a neural network's hidden layers help it recognize more complex patterns by combining information from earlier layers.

In the real world, neural networks do amazing things. They can recognize faces in photos, translate languages, predict medical conditions, and even create art. But they're not magical solutions. These systems have significant limitations. They need massive amounts of training data, and they can accidentally learn and repeat biases present in that data. Sometimes they work like a black box, deciding in ways we can't easily explain.

Take image recognition as an example. A neural network might become great at identifying dogs in photos. But show it a slightly unusual dog breed or a dog in an unexpected position, and it might get confused. Therefore, human oversight remains crucial. The neural network is a powerful tool, but it still needs human guidance and understanding.

For anyone curious about technology, neural networks are a fascinating part of artificial intelligence. They show how machines can learn and change in ways that almost seem magical. However, they also remind us that actual intelligence is complicated and involves more than just spotting patterns.

Now, understanding how AI learns is easier when you can see the process in action. Ready?

## AI Pattern Recognition Experiment

This hands-on experiment will give you a tangible experience of machine learning while having some fun through Google's Quick, Draw! game.

### What You'll Need

- A computer or tablet with internet access
- Web browser
- 15-20 minutes of uninterrupted time
- A willingness to draw (even if you're not an artist!)

### Instructions

1. Access the Tool
2. Open your web browser
3. Visit https://quickdraw.withgoogle.com/
4. Click "Start" to begin the game

### Understand the Basics

- The game will give you a word (like "cat" or "tree")
- You have 20 seconds to draw that object
- The AI will try to guess what you're drawing
- If it guesses correctly, you move to the next challenge

**Tips for the Experiment**

- Don't worry about being a great artist
- Try drawing the same object in multiple styles

As you play, pay close attention to:

- How quickly the AI recognizes your drawing
- What elements seem to help or confuse the AI
- The progression of how it guesses your drawings
- Notice how the AI responds to different drawing styles

After playing the game, consider these questions:

1. What types of drawings did the AI recognize most easily?
2. Were there any objects that consistently confused the AI?
3. What do you think made some drawings easier to recognize than others?
4. Did you notice any improvement in recognition as you drew more carefully?
5. How is this similar to how a child might learn to recognize objects?
6. What limitations did you observe in the AI's recognition?
7. How might these limitations impact real-world AI applications?
8. What surprised you most about how the AI processed your drawings?

This exercise turns machine learning from a confusing idea into something you can actually understand. As you play the game and think about what happened, you'll see how AI figures out how to sort and recognize information. It's like learning by doing instead of just reading about it.

TRY IT YOURSELF

## Identify AI in Your Everyday Life

Let's put your new knowledge to work by identifying AI systems you already use in your daily life.

Take a moment to consider your typical day. From the moment you wake up to when you go to sleep, you likely interact with multiple AI systems, often without realizing it.

Does your email have a spam filter? That's AI working to classify messages.

Do you use a navigation app like Google Maps or Waze? Those use AI to predict traffic patterns and suggest optimal routes.

Have you noticed how your smartphone's keyboard predicts what you're going to type next? That's a form of AI called a language model.

What about content recommendations on Netflix, Spotify, or YouTube? These systems watch what you like and predict more things you might want to watch or listen to.

Even more subtle: if you've used a smartphone camera in the last few years, AI is probably enhancing your photos in real-time, adjusting for lighting conditions and automatically focusing on faces.

For a more interactive exercise, try this: for one day, keep a simple log of every digital service you use. Then ask yourself:

*"Is this service likely using AI to personalize, predict, or automate something for me?"*

AI already plays a pervasive role in your daily routines, which might surprise you.

Here's another way to think about it: AI shows up in tasks that are:

- Repetitive but require some judgment
- Based on recognizing patterns in data
- Personalized to individual preferences
- Adaptive to changing conditions

The goal of this exercise isn't to make you paranoid about AI's presence in your life. Far from it. Rather, it's helping you recognize AI isn't some futuristic concept; it's already a practical technology enhancing many of the services you rely on.

Understanding this reality is the first step toward demystifying AI and seeing it as a tool rather than a mysterious force. Those examples show how AI works in practice. To make sure we understand the main ideas, here's a table to recap them.

| | **Definitions** | **Examples** |
|---|---|---|
| **Artificial Intelligence** | Creating computer systems that can perform tasks that typically require human intelligence | Expert Systems<br>Rule-based Automation |
| **Machine Learning** | Systems that learn from data and improve with experience | Recommendation Systems<br>Predictive Analytics |
| **Deep Learning** | Multi-layered neural networks that process complex patterns | Image Recognition<br>Natural Language Processing |

## MOVING FORWARD

As we wrap up this foundational chapter, you now have a clearer picture of what AI is and isn't. You understand its key components, how it works at a high level, and you've identified its presence in your daily life.

In the next chapter, we'll explore AI in your daily life in greater detail, looking at specific applications from virtual assistants to content recommendations, and even how AI is changing healthcare for individuals.

Remember, the goal isn't to make you an AI engineer or data scientist. It's giving you the knowledge and confidence to understand, use, and talk about AI in practical ways. Because AI literacy isn't just for tech experts. It's becoming an essential skill for everyone.

### References

1. *McCarthy, J., Minsky, M. L., Rochester, N., & Shannon, C. E. (1955). A proposal for the Dartmouth summer research project on artificial intelligence. AI Magazine, 27(4), 12-14.*

2. *Manyika, J., Lund, S., Chui, M., Bughin, J., Woetzel, J., Batra, P., Ko, R., & Sanghvi, S. (2017). Jobs lost, jobs gained: Workforce transitions in a time of automation. McKinsey Global In*

"Technology, like art, is a soaring exercise of the human imagination."

Daniel Bell

# CHAPTER 02

# AI in Your Daily Life

DID YOU KNOW YOU PROBABLY INTERACTED with at least a dozen AI systems this morning? I mean it. Think about your day so far. A smart speaker woke you up. Your phone's map showed the best route to work. Your email sorted important messages automatically. AI is everywhere, working so smoothly that you barely notice it.

This is often how major shifts in technology unfold. They begin with a burst of excitement, curiosity, and sometimes controversy. Over time, though, those once-revolutionary changes settle into daily life. Electricity probably felt like a miracle when it first arrived in homes, lighting up rooms with the flick of a switch. Today, we rarely think about it unless there's a power outage or our devices stop charging.

Artificial intelligence is somewhere in the middle of this journey. Some parts still leave us wide-eyed, like lifelike voice assistants or art generated in seconds. Other AI features, like predictive text or spam filters, are already so routine that we barely notice them working in the background. Now is a good time to take a closer look and see where AI is quietly shaping your everyday experiences, and where it might be adding a few new twists.

## Virtual Assistants and Smart Home Devices

*"Hey Siri, what's the weather today?"*

*"Alexa, add milk to my shopping list."*

*"OK Google, play my morning playlist."*

Sound familiar? These voice commands show how AI has moved into our homes. What started as basic voice tools that could barely understand us have become smart helpers. Now they can control your home, answer questions, suggest things you might like, and even try to make you laugh (though their jokes are pretty terrible).

So, how do these assistants actually work? At their core, virtual assistants rely on a mix of AI technologies, including:

- Speech recognition converts your spoken words into text.
- Natural language processing (NLP) interprets what you're asking.
- Machine learning helps the system improve over time, learning your preferences, speech patterns, and common requests.

A few years ago, I helped a friend set up a smart home system in his new house. He loved technology but had serious mobility challenges, so being able to control lights, temperature, and entertainment with his voice made a tremendous difference. I'll never forget the look on his face when he first said, "Alexa, I'm home," and the lights turned on, his favorite music started playing, and the thermostat adjusted to his ideal setting. He told me,

*"It's like having a butler who never needs a day off."*

I learned something crucial about AI that day: we often get caught up in big technological breakthroughs, but the real magic happens when technology makes daily life smoother. For my friend, these small automations made a vast difference in how he could move around and live more independently.

Smart home devices have come a long way since then. Now, they can tell who's talking, learn your daily routines without you having to program everything, and even spot something weird happening at home (like a water leak or unusual activity when you're not around).

But this convenience has drawbacks, especially for privacy. These devices are always listening for words like "Hey Siri" or "Alexa." Companies say they record only after hearing these words, but some people worry others could misuse or hack them.

I once talked to an IT security professional who kept his smart speaker in a drawer whenever he wasn't using it. When I asked if that was over the top, he laughed and said,

> *"Fifteen years ago, the idea of willingly putting an always-on microphone in your home would have sounded ridiculous. Now we actually pay to do it."*

It's a fair point. The normalization of surveillance has happened gradually, with each small convenience making the next privacy concession easier to accept. I'm not suggesting you should unplug all your devices. I certainly haven't. But awareness of these trade-offs helps us make more informed choices about which conveniences are worth the potential costs.

If you're concerned about privacy but still want the benefits of smart assistants, consider:

- Using the mute button when you're having sensitive conversations
- Regularly reviewing and deleting your voice history (most assistants allow this)
- Being selective about which features you enable and which third-party skills or services you connect

I often ask my clients a simple question: "If you could only keep three things your smart speaker does, what would they be?" Most people are shocked to realize how few features they actually use, even though these devices can do tons of things. With technology, sometimes doing less is better, especially when it helps protect your privacy.

## AI in Social Media and Content Recommendations

*"How does Netflix always know what I want to watch?"*

*"Why do I keep seeing ads for products I was just talking about?"*

*"My TikTok feed seems to know me better than my spouse does!"*

I hear these comments all the time: people wondering how websites and apps seem to know exactly what they want. What you see online isn't random. Smart computer systems carefully choose the content in your social media, streaming services, shopping sites, and search results.

How do they do it? It's pretty clever. These systems watch what you click on, like, and spend time with. Then they compare your choices to millions of other people's behaviors. From that, they guess what might catch your eye next.

The result? Websites and apps now feel like they know you almost too well. I call this the "digital mirror" effect. Our devices show us exactly what we want to see, creating cozy bubbles that can actually limit what we learn.

I saw this up close while working with a local news website struggling to get readers. When we looked at their content, we found something surprising. Their best, most carefully researched stories weren't getting nearly as many clicks as their bold, attention-grabbing headlines that stirred up powerful emotions.

"Our system loves content that gets big reactions," the digital editor told me. "Calm, balanced stories just don't grab people's attention."

This reveals a hidden problem with how online content spreads. Stories that make people angry, shocked, or scared travel much faster than calm, thoughtful reporting. The people who design these systems didn't mean to create division or spread false information. But by focusing only on what gets the most clicks and shares, they accidentally created a system that rewards extreme content.

ETHICAL SPOTLIGHT

**Algorithmic Filtering and Echo Chambers**

Recommendation algorithms show you content you are most likely to enjoy. However, this kind of filtering has a drawback. It can create "filter bubbles" and "echo chambers," where you mostly see content that supports your existing beliefs.

A study from the University of Exeter found social media can create "echo chambers," where people mainly interact with others who already agree with them. This can make their beliefs even stronger.[1] When recommendation systems focus more on keeping people engaged than on offering variety, they can unintentionally increase biases and limit exposure to new ideas.

That's why it's important to stay aware of how these systems work and make a conscious effort to seek out different perspectives.

To prevent this, ethical AI design should include safeguards. Platforms can add features that promote a mix of information, even while tailoring content. For example:

- **Content variety nudges:** Algorithms can occasionally surface content with opposing viewpoints or from different sources.
- **Transparency features:** Apps can display why they recommend certain content, giving users more insight into the filtering process.

As a user, you can take steps to avoid getting stuck in a content bubble:

1. Periodically reset your recommendations on platforms like YouTube, Netflix, or Spotify.
2. Intentionally engage with diverse content, even if it doesn't align with your usual preferences.
3. Use private browsing or separate accounts for exploring new topics versus regular consumption.
4. Stay aware that algorithms filter what you see, and the online world they present may not reflect the full range of perspectives.

### Taking Control of Your Digital Experience

Algorithms aren't all bad. In fact, they help businesses connect with people who genuinely care about their content. For users, they can surface interesting information that would otherwise get buried in the endless flood of online content.

The key is to stay mindful of how these systems shape your experience. By making small, intentional choices, you can keep your information diet balanced and avoid getting stuck in a digital echo chamber.

## AI-Powered Tools for Personal Productivity

Now, let's move from how AI affects what we see and read to how it can help us be more productive and organized. This is where AI goes from working in the background to becoming a helpful assistant, using tools that boost what we can do.

**Writing Help**

Whether you're crafting an important email, working on a report, or just trying to find the right words, AI writing tools have exploded in capability and accessibility over the past few years.

Writing tools like Grammarly have become much smarter than just catching spelling mistakes. Now they help you write better by suggesting ways to improve your tone, make your writing clearer, and keep readers interested. It's like having a helpful editor right in your computer.

Text generation tools like ChatGPT, Claude, or Google's Bard can help overcome writer's block, suggest alternative phrasings, summarize complex documents, or even draft complete responses based on minimal input.

While researching for this book, I spoke with the owner of a small marketing agency that started using AI writing tools. At first, she worried these tools might replace her creative team or make their work feel less unique.

A few months later, she shared a surprising discovery.

> *"Our writers aren't doing less or getting lazy. They're actually creating better content and spending more time on big creative ideas while letting AI handle the routine writing tasks."*

The best AI tools help people work better, not replace them. They take care of boring, repetitive tasks so humans can focus on the creative and thoughtful parts of their work.

But these tools aren't perfect. Sometimes they write something that sounds good but isn't actually true. They can miss important cultural details and create boring content. Think of them as helpful assistants, not replacements for human thinking and skill.

This reminds me of a conversation with a professor who caught students using AI-generated essays. "The worst of it wasn't that they just used AI to write their essay," he told me. "It's that they used it badly. They accepted whatever it produced without critical evaluation or personalization."

Here's the key: treat AI writing tools like a helpful partner. Give them explicit instructions, double-check the facts, and always add your own personal touch to make the writing truly yours.

## Organizing and Improving Photos

Many people have thousands of digital photos spread across devices, cloud storage, and social media. AI can help by sorting, finding, and improving pictures with smart organization and editing tools. Modern photo apps can automatically:

- Categorize images by recognizing people, places, objects, and events
- Identify your best shots based on technical quality and composition
- Apply sophisticated enhancements tailored to specific image types
- Create collections and highlight meaningful moments without manual sorting

A photographer friend recently showed me how he uses AI tools to speed up his workflow. "Ten years ago, I spent hours sorting and tagging images by hand," he said. "Now, AI handles the initial organization, so I can focus on creative editing and working with clients."

What feels like magic is really just clever computer programs. They've studied tons of photos, so now they're great at finding patterns, picking out faces and objects, and even guessing how good a picture looks. Some can figure out which shots you'll probably love the most by sensing the emotions or beauty in an image.

Of course, there are still occasions when AI gets it hilariously wrong. My phone recently created an automatically generated album called "Fun Times" that consisted mostly of photos I'd taken of receipts from different places I traveled to for tax purposes. Not quite the nostalgic trip it had in mind, I think.

When AI gets things wrong, it reminds us of something crucial. AI rocks at sorting through loads of data and spotting patterns, but it can't beat human judgment for understanding the big picture. The smartest way to use AI is to team up with it. Let it do the grunt work of organizing, then you step in to decide what's actually important and why it matters.

## Email Management and Scheduling

Email overload is a modern plague. According to a McKinsey analysis cited in the Harvard Business Review, the average professional receives about 120 emails per day.2 Trying to stay on top of that mess can eat up hours of time you could spend getting actual work done. But now AI is jumping in to help wrestle your inbox under control.

- Smart categorization systems automatically sort messages by importance and type
- Suggested replies offer one-click responses to common questions
- Follow-up reminders ensure important messages don't fall through the cracks
- Writing assistants help draft effective emails more quickly

The American Bar Association's 2024 Legal Technology Survey Report shows that more law firms are using artificial intelligence (AI) to work more efficiently. AI use in the legal field jumped from 11% in 2023 to 30% in 2024, with larger firms adopting it faster. While the report does not give exact numbers on time saved from AI-powered email management, 54% of those surveyed said AI tools made them more efficient, especially with document review and management. The report also found that 73% of firms use cloud-based legal tools, and more firms are using AI for discovery tasks. This shift allows attorneys to spend more time on legal work instead of routine tasks.3

AI can also make scheduling meetings a lot less painful. No more endless emails trying to find a time that works for everyone. Tools like Calendly, x.ai, and Clara can look at people's calendars, learn what times they like to meet, and even set up meetings by sending emails that sound like a real person wrote them.

*One person I interviewed described her AI scheduling assistant as "the one digital tool I'd never give up. It's saved me from hundreds of 'how about Tuesday at 2? No wait, that doesn't work for me' email chains."*

These AI tools can save you a ton of time on boring admin tasks, but you need to be smart about how you use them. Emails can contain private info, so an AI making a mistake or sending the wrong auto-reply could cause problems. The key is to have a human double-check the important messages while the AI takes care of the everyday stuff.

## Healthcare Applications for Individuals

You might hear about AI doing amazing things in healthcare, like figuring out what's making someone sick or finding new medicines. But AI is also working hard to help regular people stay healthy in their everyday lives. There are apps and gadgets you can use right now that monitor your health, spot problems early, and help you make smarter choices about your well being.

### Health Monitoring and Predictive Tools

Wearable devices like smartwatches and fitness trackers now incorporate increasingly sophisticated AI to:

- Detect irregular heart rhythms that might show atrial fibrillation
- Monitor sleep patterns and identify potential sleep disorders
- Track activity levels and suggest personalized fitness goals
- Detect falls and automatically alert emergency contacts

A cardiologist I spoke with for this book shared a story about a patient whose smartwatch picked up an irregular heartbeat during his usual morning routine.

> *"He had no symptoms yet," she said. "But because we caught it early, we could start treatment before it turned into a serious cardiac event. These devices aren't perfect, but they're changing the game by providing continuous, passive monitoring that can catch issues traditional screenings might miss."*

Stanford Medicine found that smartwatches and other wearables using AI could spot irregular heartbeats with 97% accuracy, almost as well as professional EKG tests.4 These devices can't replace a doctor, but they help detect heart problems early and track useful health data.

If you have an ongoing health problem, AI-powered apps can make it easier to keep track of your symptoms, remember to take your medicine, and notice things in your environment that might make your condition worse. Some of these apps are so smart they can even predict when you might have a flare-up or complication before it happens. That way, you can take steps to prevent problems instead of just reacting to them.

A friend with Type 1 diabetes recently showed me how her continuous glucose monitoring system uses AI to predict potential low blood sugar events before they happen. "It's given me peace of mind," she said. "Especially at night, knowing I'll get an alert before my levels become dangerous."

### Personalized Wellness Recommendations

Beyond monitoring specific health metrics, AI is increasingly helping individuals optimize overall wellness through personalized recommendations:

- Nutrition apps that suggest meal plans based on health goals, preferences, and dietary restrictions
- Fitness programs that adapt workouts based on performance, recovery, and equipment
- Mental health tools that offer customized meditation or cognitive-behavioral therapy exercises
- Sleep optimization systems that analyze patterns and suggest improvements

The power of these applications comes from their ability to adapt continuously based on your data, creating truly personalized wellness plans rather than one-size-fits-all approaches.

While researching this book, I tested several AI fitness apps, and what stood out most was how quickly they adapted to my feedback. When I logged a certain exercise hurt my knee, the app immediately adjusted my future workouts to avoid similar movements while still working the same muscles.

Like other AI tools, these fitness apps work best as partners, not decision-makers. They can analyze patterns in your data and offer helpful suggestions, but you know your body best.

A nutritionist I interviewed put it well: "AI can analyze patterns in your eating habits, sleep, and activity levels that you might miss. But it doesn't know how you feel or what motivates you. The most successful users treat these tools as informed advisors, not infallible oracles.

TRY IT YOURSELF

## Automate a Simple Task Using AI

Now for the fun part. Let's put your new knowledge to work by automating a simple, everyday task using consumer AI tools that require zero coding skills.

Here's a practical project that combines several AI capabilities we've discussed: creating an automated email management system to reduce inbox stress.

**What you'll need:**

- Gmail or Microsoft Outlook (both have built-in AI features)
- 30 minutes to set up

### Step 1: Analyze Your Email Patterns

Start by reviewing your inbox and identifying recurring patterns in the emails you receive. Look for:

- Newsletters or updates you regularly read but don't need immediate attention.
- Common request types that require similar responses (e.g., meeting requests, project updates, etc.).
- Time-sensitive messages that need quick action, such as reminders or urgent emails from clients or colleagues.

### Step 2: Create Smart Filters and Categories

Next, set up automated organization rules using the built-in filter and categorization tools in Gmail and Outlook.

**In Gmail:**

1. Go to Settings (click the gear icon in the top-right corner).
2. Select See all settings.
3. Navigate to the Filters and Blocked Addresses tab.

4. Click on Create a new filter.
5. Define the filter criteria:

   ◦ Newsletters: Type the common sender's address (e.g., "newsletter@company.com") or keywords (e.g., "unsubscribe") into the filter fields. Then, click Create filter.

   ◦ Receipts and Confirmations: Add common phrases like "receipt," "order confirmation," or "invoice" to categorize these emails automatically into a Purchases folder.
6. To automatically sort these emails into categories or labels, check the option to Apply the label and choose or create a label (e.g., "Reading," "Purchases," "Work").
7. For high-priority messages from key contacts, select Never send it to Spam and create a Starred category.

You can find a more detailed step-by-step information on creating filters in Gmail here:

https://bit.ly/4bRNpst

**In Outlook:**

1. Open Outlook and go to Home.
2. Click on Rules in the ribbon, then select Manage Rules & Alerts.
3. In the Rules and Alerts dialog, click New Rule.
4. Choose Apply rule on messages I receive.
5. Set up rule conditions like:

   ◦ Newsletters: Choose keywords such as "newsletter" or the sender's email address to filter them into a specific folder.

   ◦ Receipts and Confirmations: Create a rule for moving emails with keywords like "receipt," "order," or "invoice" into a Purchases folder.
6. Select actions like Move to Folder and choose the folder you want emails to go to, such as "Reading" or "Purchases."

More detailed step-by-step information for setting up rules in Outlook: https://bit.ly/41CAOob

### Step 3: Set Up Templates for Common Responses

Creating templates for frequently used responses will save time and keep your communication consistent. You can set these up in Gmail and Outlook:

**In Gmail:**

1. Open Gmail and click on Settings (gear icon).
2. Go to See all settings > Advanced.
3. Scroll down to Templates and select Enable.
4. Once enabled, compose an email that you use frequently (e.g., meeting scheduling or project updates).
5. Click on the three-dot menu in the bottom-right corner of the email window and select Templates > Save draft as template > Save as new template.

For more detailed information:

https://bit.ly/4iBeLoS

**In Outlook:**

1. Open Outlook and compose a new email with your template content (e.g., meeting invites, responses to common inquiries).
2. Click on the Insert tab, then select Quick Parts > Save Selection to Quick Part Gallery.
3. Name your template and choose a category for easy access.
4. To use the template, simply start a new email, click Quick Parts from the Insert tab, and select your saved template.

For more detailed information:

https://bit.ly/3Fyvgn7

### Step 4: Enable Smart Features

Both Gmail and Outlook offer AI features to help speed up your email writing process.

**In Gmail:**

1. Go to Settings > See all settings > General.
2. Scroll to the Smart Compose section and select Enable.
3. Turn on Smart Reply to get quick response suggestions when you open an email.

More detailed step-by-step information for enabling Gmail's smart features: https://bit.ly/4iAVBj4

**In Outlook:**

1. Go to File > Options > Mail.
2. Scroll down to Text Predictions and select Use text predictions.
3. This will help you complete sentences and suggest words as you write emails, saving time.

More detailed step-by-step information for enabling Outlook's text predictions: https://bit.ly/3FAMm3W

### Step 5: Set Dedicated Email Time

Set specific times in your calendar to check and respond to emails. This dedicated time will help you manage the messages more effectively, allowing you to focus on high-priority emails while automation takes care of the rest.

**Tip:** Schedule time twice a day (e.g., once in the morning and once in the afternoon) to check and respond to emails based on their priority.

### Step 6: Monitor and Refine

After a week of using your new automated system, take some time to evaluate how it's working. Check if the filters are sorting emails correctly, if templates are saving you time, and if smart features are helping you respond quicker. Make any necessary adjustments to improve the system's efficiency.

**Tip:** Share your thoughts on Gmail or Outlook's smart features. They get better the more you use them and the more feedback you give.

Now, with your automated email management system in place, you can spend less time organizing and responding to emails, and more time focusing on what matters most!

This project shows how AI tools can lighten your mental load and save time without needing technical skills. Once you get the hang of it, try automating other repetitive tasks in your daily routine.

## MOVING FORWARD

AI is already a big part of our daily lives, shaping what we see online and powering tools that help us stay productive and healthy. But this is just the beginning of a much bigger shift in how technology supports human abilities.

What makes this moment exciting is that AI is still new enough for people to shape how they use it. Unlike past tech shifts led by big companies, today's AI tools are easy to access, letting anyone experiment and tailor them to their needs.

In the next chapter, we will explore how anyone can use no-code AI tools. You will learn how to create AI applications with no coding, making things possible that once seemed out of reach.

Remember, the goal isn't to use AI for everything, but to thoughtfully integrate it where it truly adds value to your life, enhancing your capabilities while respecting your privacy and autonomy.

## References

1. *Williams, H. T., McMurray, J. R., Kurz, T., & Lambert, F. H. (2015). Network analysis reveals open forums and echo chambers in social media discussions of climate change. Global Environmental Change, 32, 126-138.*

2. *Bailey, E., & Konstan, J. (2019). How to spend way less time on email every day. Harvard Business Review.*

3. *American Bar Association. (2024). 2024 Legal Technology Survey Report. ABA Journal. Retrieved from https://www.abajournal.com*

4. *Tison, G. H., Sanchez, J. M., Ballinger, B., Singh, A., Olgin, J. E., Pletcher, M. J., Vittinghoff, E., Lee, E. S., Fan, S. M., Gladstone, R. A., Mikell, C., Sohoni, N., Hsieh, J., & Marcus, G. M. (2018). Passive detection of atrial fibrillation using a commercially available smartwatch. JAMA Cardiology, 3(5), 409-416.*

**The best AI tools help people work better, not replace them.**

"The most profound technologies are those that disappear. They weave themselves into the fabric of everyday life until they are indistinguishable from it."

Mark Weiser

# CHAPTER 03

# No-Code AI: Power Without Programming

REMEMBER WHEN CREATING A WEBSITE REQUIRED learning HTML, CSS, and JavaScript? When building a mobile app meant mastering Swift or Java? For most of us, these technical barriers kept us firmly in the "user" category rather than the "creator" camp.

But then something changed. Website builders like Wix and Squarespace came along, making it easy for anyone with an idea to create a website with no coding required.

We're witnessing the same transformation with artificial intelligence right now.

Recently, using AI meant you needed an advanced degree in computer science, coding skills in languages like Python, and experience with frameworks like TensorFlow or PyTorch. These technical barriers made AI out of reach for most people.

Not anymore.

No-code AI platforms are making advanced technology available to everyone. Now, even without coding skills, you can build AI applications that would have required a team of engineers just five years ago.

In this chapter, we'll explore the world of no-code AI: what it is, how it works, why it matters, and most importantly, how you can use it today without writing a single line of code.

## Overview of No-Code Platforms

The term "no-code" might sound like a stretch. How can you build advanced AI applications without writing code? The key is a shift in how these platforms handle software development.

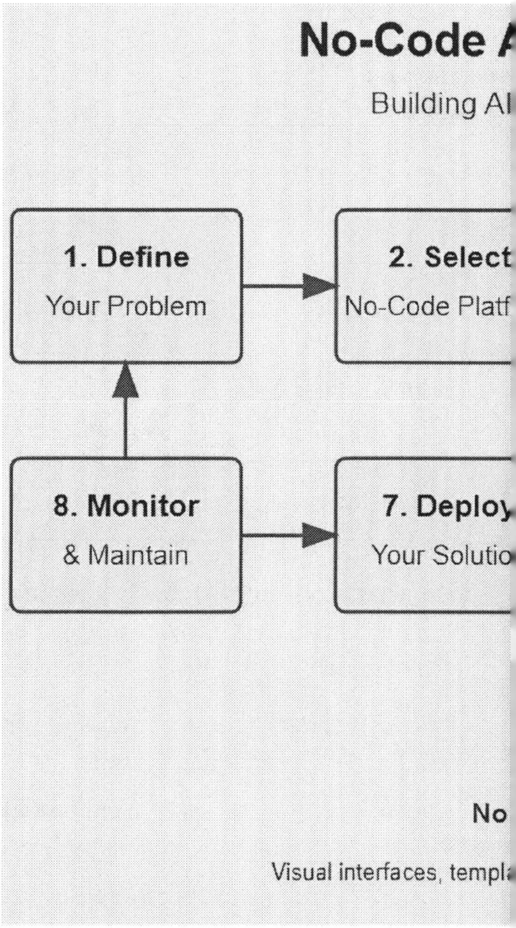

Traditional programming means giving the computer step-by-step instructions for every task. No-code platforms work differently. Instead of writing code, you simply define what you want to achieve, and the system takes care of the technical details behind the scenes.

Think of it like building a house. One way is to construct everything from the ground up, piece by piece. The other is using prefabricated parts that fit together easily. Both methods get you a solid, functional home, but one takes far less technical skill.

### Popular Tools and Their Capabilities

The no-code AI space is growing fast, with new tools popping up almost every month. Here are some of the most beginner-friendly options available today:

**Visual AI Builders**

Platforms like Obviously AI, Lobe (Microsoft), and CreateML (Apple) let you build machine learning models using simple drag-and-drop interfaces. You provide the data, choose what you want to predict, and the platform takes care of training the model.

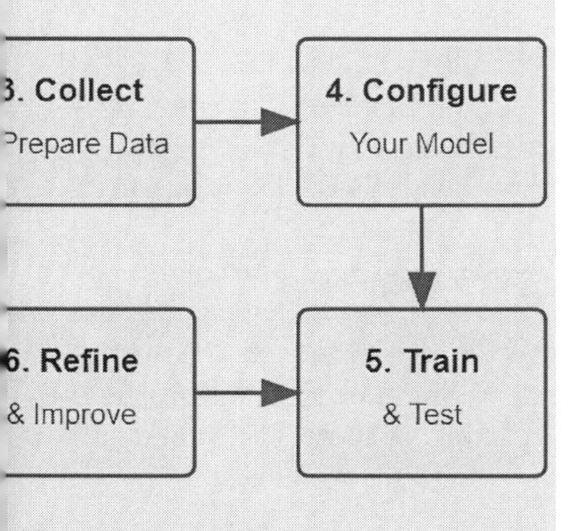


**AI Task Automation**

Tools like Zapier, Make (formerly Integromat), and IFTTT now include AI features. They help connect apps and automate tasks, using AI for things like analyzing emotions in text, recognizing images, and understanding language.

**Natural Language Interfaces**

The easiest way to start with no-code AI is through conversational tools like ChatGPT, Claude, and Bard. You can describe what you need, and they can generate code, analyze data, create content, and more—no technical skills required.

## User-Friendly Interfaces

The core of any no-code AI platform is its interface. The success of these tools depends on how well they simplify complex technical processes into easy-to-use interactions. Most platforms follow a few common design approaches:

### Visual Programming

Many use drag-and-drop interfaces where you connect different components to build workflows. This approach looks like a flowchart, making it easy to see and adjust the logic without writing code.

### Guided Wizards

Other platforms use step-by-step wizards that walk you through the process of creating an AI application. These are effective for beginners as they provide structure and explanation at each stage.

### Templates and Pre-Built Solutions

Many platforms provide ready-made templates for common tasks, so you don't have to start from scratch. These templates make it easier to get started while still allowing for customization.

## Benefits of No-Code AI

The rise of no-code AI isn't just a technical trend. It's changing who can build and use AI, making it more accessible to everyone. This shift comes with significant advantages:

### Quicker Development

Traditional AI projects take months because they involve collecting data, preparing it, training models, testing, and maintaining them. No-code AI simplifies this process, letting users create and test solutions much faster.

No-code platforms compress this timeline dramatically. Tasks that once took quarters now often take just days or even hours.

### Cost-Effectiveness

The economics of AI development are being reshaped by no-code platforms. Traditional AI development comes with a high price tag. Hiring a data scientist in the US costs around $120,000 per year. On top of that, there are expenses for infrastructure, software licenses, development tools, and ongoing maintenance. For many smaller organizations, these costs make AI out of reach.

No-code platforms make AI more affordable with subscription-based pricing. Most offer tiered plans based on usage, with entry-level options starting at under $100 per month. Even at the enterprise level, no-code solutions are usually much cheaper than building AI from scratch.

## AI for Everyone

No-code AI makes advanced technology accessible to people who couldn't use it before.

Business experts who know the challenges but lack coding skills can now build their own AI solutions. Creatives and designers can explore AI-powered workflows without technical barriers. Small business owners can compete with larger companies by using AI tools. Educators can create interactive learning experiences tailored to their students. Even researchers from non-technical fields can apply AI to their work without needing to learn to program first.

? ETHICAL SPOTLIGHT ?

### Democratization vs. Responsibility

No-code AI tools make powerful technology easy for anyone to use. However, this also makes it easier to create systems that can affect the real world. With this comes important questions. Should all AI projects go through an ethical review? How can we make sure no-code developers understand the risks of what they build? Even without coding skills, creators should think about the misuse of their AI or unexpected problems.

# How to Get Started with No-Code AI

No-code AI is becoming easier to use, but the many options can feel overwhelming for beginners. Here's a simple way to get started:

## Choosing the Right Platform for Your Needs

Finding the best no-code AI platform starts with understanding your needs. Here are the key factors to consider:

### Define Your Problem

Different platforms specialize in different AI tasks.

- Data predictions or classifications, try Obviously AI, MonkeyLearn, or BigML.
- Image and video analysis, explore Teachable Machine, Lobe, or Roboflow.
- Text processing, Levity or MonkeyLearn are great options.
- Use AI-powered automation like Zapier, Make, or Workato.

### Consider Your Technical Comfort Level

Some platforms are fully visual, allowing you to build AI models with drag-and-drop tools. Others might require simple formulas or scripting. It is important to be honest about your comfort level when choosing a platform. Selecting one that is too complex can lead to frustration and stalled progress.

### Assess Your Data Needs

- Some platforms need large datasets, while others work with smaller data or pre-trained models.
- The format matters too. Do you have spreadsheets, images, or text? Your data type will help determine the right tool.

### Factor in Your Budget

- Many platforms offer free tiers, perfect for minor projects and experimentation.

- Costs usually scale based on usage, so check long-term pricing before committing.

For businesses unsure about AI, I often suggest starting with free or low-cost options. This allows them to experiment without a big financial commitment.

## Resources for Learning

Once you've chosen a platform, the next step is learning how to use it effectively. Luckily, there are plenty of resources to help:

### Platform-Specific Learning Materials

Most no-code AI platforms offer built-in education tools, including:

- Interactive tutorials walking through common use cases
- Step-by-step documentation and knowledge bases
- Video demos and webinars
- Template galleries with pre-built AI models

These are the best place to start since they focus on the exact tools you'll be using.

### Online Courses and Tutorials

If you want to dive deeper, platforms like Udemy, Coursera, and YouTube offer no-code AI courses. Specialized sites like No-Code MBA provide AI-focused modules for business users.

### Communities and Forums

Some of the best learning happens through discussions with other users. Look for:

- Platform-specific communities on Discord, Slack, or forums
- No-code groups like Makerpad and NoCode HQ
- Reddit discussions in r/nocode and r/MachineLearning
- Skool classrooms focused on no-code AI

I've seen beginners solve problems in minutes just by asking a question in this space, something they might have spent weeks figuring out alone. The collective knowledge in these communities is invaluable.

## Common Pitfalls to Avoid

Even with user-friendly interfaces, there are common mistakes beginners make when starting with no-code AI. Being aware of these pitfalls can save significant time and frustration.

### Unrealistic Expectations

No-code platforms make AI easier to use, but they are not magic. I have seen users give up on promising projects because their expectations were too high.

To avoid this, set realistic expectations by learning what your chosen platform can and cannot do. Start with simpler projects and gradually take on more complex ones as you gain experience.

### Poor Data Quality

The old computing adage "garbage in, garbage out" applies doubly to AI systems. No-code platforms can automate many aspects of AI development, but they can't transform bad data into good results.

### Making Things Too Complicated

A common mistake is adding unnecessary complexity. No-code platforms are very flexible, but sometimes users build overly detailed solutions when a simpler one works just as well.

Start with the simplest viable solution that solves your core problem. Once you have a solid foundation, you can always add complexity if needed.

## Success Stories from Non-Technical Users

Guidelines are helpful, but real-world examples can teach us the most. Let's look at how non-technical users have successfully built AI solutions using no-code platforms.

### Increasing Student Retention at Ivy Tech Community College

Ivy Tech Community College, Indiana's largest public college, had trouble keeping students from dropping out. They wanted to spot at-risk students sooner, but didn't have data science experts to build a traditional AI system.

Using the no-code platform Civitas Learning, staff without technical backgrounds created an early warning system. It analyzed student data to predict dropouts and uncovered surprising patterns, like certain course combinations and class schedules that increased the risk of students leaving.

The impact was significant. Ivy Tech increased its retention rate by three percentage points in the first year, keeping hundreds of students enrolled. According to their case study, this improvement translated to roughly $1.5 million in retained tuition revenue and better student outcomes.[1]

### AI-Powered Food Waste Reduction at Migros Supermarkets

Swiss supermarket chain Migros wanted to cut down on food waste by improving how they predicted demand and managed inventory. Store managers used to rely on manual estimates to stock fresh food, which often led to either too much waste or not enough products on the shelves.

To fix this, Migros teamed up with Blue Yonder, a company specializing in AI-powered supply chain solutions. They introduced a machine learning system that looked at past sales, weather, promotions, and local events to predict the right amount of fresh food to stock.

The results were impressive. According to Migros' reports, stores using AI-based forecasting saw major improvements.

- Reduced food waste by 20% within the first year.
- Decreased out-of-stock incidents by 15%, improving customer satisfaction.

- Increased overall profitability by optimizing inventory levels.

"The AI solution allowed us to adjust stock levels dynamically, reducing both food waste and lost sales opportunities," said Cédric Moret, CIO of Migros Group. "What's remarkable is that the system continuously learns and adapts to shifting customer behaviors, such as seasonal demand and unexpected events."2

Migros continues to expand AI-driven solutions across its supply chain, reinforcing its commitment to sustainability and efficiency.

### Automating Marine Conservation Research

The conservation group Oceans Initiative needed to study thousands of underwater recordings to track endangered orcas in the Pacific Northwest. Before, they had to either have experts listen to hours of audio by hand or hire programmers to create special algorithms.

Dr. Erin Ashe, a marine biologist with no coding experience, turned to Google's Teachable Machine, a no-code AI platform. She and her team trained a model to recognize orca calls and differentiate them from other marine sounds. Instead of taking months or years, they completed the process in just weeks.

> *"The model now successfully identifies over 90% of orca vocalizations in our recordings," reported Ashe. "What's remarkable is that we built and refined it ourselves without needing to hire developers or learn programming."3*

The team now processes in hours what previously took weeks, allowing them to track population movements in near real-time and better inform conservation efforts.

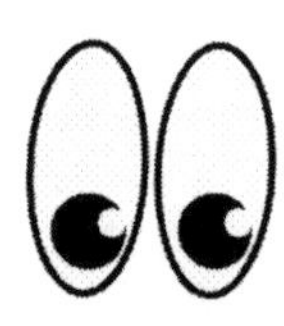

Want to try it yourself? While we won't be working with orca recordings, this hands-on project will show the power of no-code AI using the same platform Oceans Initiative used. In this exercise, you will create your own AI image recognition system using Google's Teachable Machine, a free browser-based tool that lets anyone build machine learning models without writing a single line of code.

## Create a Custom AI Image Recognition Model

Ever wondered how Facebook detects faces in photos or how your phone groups pictures by what's in them? In this exercise, you'll build your own AI image recognition system using Google's Teachable Machine

### Tools Needed:

Computer with webcam and internet connection

### Step 1: Set Up Your Project

1. Open your web browser and go to teachablemachine.withgoogle.com
2. Click the "Get Started" button
3. Select "Image Project" from the options
4. Choose "Standard image model"

You'll see a workspace with three default classes labeled "Class 1," "Class 2," and "Class 3." Think of these as categories or buckets you're going to teach the AI to recognize.

### Step 2: Plan Your Recognition Categories

For this exercise, we'll create a gesture recognition model. Decide on three distinct hand gestures, such as:

- Thumbs up
- Peace sign
- Open palm

Choose different gestures if you prefer, but make sure they're visually distinct from each other.

### Step 3: Collect Training Data

1. Rename your classes by clicking on "Class 1" and changing it to "Thumbs Up" (or your chosen gesture name)
2. Repeat for the other classes, naming them according to your chosen gestures

Now let's gather training examples:

1. Select the first class ("Thumbs Up")
2. Click the webcam icon if it's not already selected
3. Position your hand in the correct gesture in front of your webcam
4. Click and hold the "Hold to Record" button while moving your hand slightly in different positions and angles
5. Capture at least 20 samples, varying the:
    - Position of your hand in the frame
    - Distance from the camera
    - Angle of your hand
    - Lighting conditions if possible

Repeat for each of your other gesture classes, capturing at least 20 samples for each.

The key to a robust model is variety in your training examples. Don't just take 20 nearly identical images. Move your hand around, change the angle, try different backgrounds, and adjust the distance from the camera. This teaches the AI to recognize the essential patterns of each gesture regardless of these variations.

### Step 4: Train Your AI Model

1. Click the "Train Model" button in the upper right corner of the screen
2. Wait as the system analyzes your images and builds a machine learning model (this typically takes 30, 60 seconds)
3. Once training is complete, you'll see a preview window where you can test your model in real time

Try performing each gesture in front of your webcam. The model will show its prediction along with confidence percentages for each class. Notice how the confidence levels change as you move your hand.

### Step 5: Test and Refine

Now it's time to see how well your model works:

1. Test each gesture several times, noting how accurately the model recognizes them
2. Try "edge cases" such as positions or angles you didn't include in your training data
3. If the model struggles with certain variations, add more training examples:
    - Click on the class that needs improvement
    - Capture more samples focusing on the angles or positions that caused errors
    - Click "Train Model" again to update your AI

Professional AI developers follow this iterative improvement process by testing, identifying weaknesses, adding more training data, and retraining the system.

Behind the scenes, the AI is not actually "understanding" what a hand or gesture is. Instead, it analyzes patterns of pixels, looking for visual features that consistently appear in one category but not in others. Therefore, diverse training examples are so important—they help the AI focus on what matters, like the shape of your hand, while ignoring things like background, lighting, or exact positioning.

### Step 6: Share Your AI Creation

Modern AI tools empower users to share and build upon them with ease. Let's save your model:

1. Click the "Export Model" button in the upper right
2. In the dialog that appears, select the "Shareable Link" tab
3. Click "Copy" to get a URL for your model
4. Save this link somewhere safe

This link allows anyone to use your trained model directly in their browser. Congratulations! You've just created and deployed your first AI application without writing a single line of code!

For an extra challenge, try building a more complex model with five or more different gestures. You could also train the AI to recognize objects like household items instead of hand gestures. Another interesting experiment is to add a "Nothing" or "Background" class with images of your environment when no gesture or object is present. This helps the model learn what not to classify, improving accuracy.

*Want to see AI image recognition in action before creating your own? Visit this demo I created on Teachable Machine to try it out and judge for yourself. I built this model using the same steps you'll follow in this exercise!*

https://bit.ly/4iplVfI

## CONNECTING TO THE CHAPTER

This exercise demonstrates several key concepts we've discussed throughout this chapter:

1. Supervised learning firsthand by providing labeled examples
2. How data quality directly affects performance
3. An iterative improvement process that's core to AI development
4. Used a powerful machine learning platform without writing a single line of code

The next time you use facial recognition to unlock your phone or search for "beach photos" in your image library, you'll understand the process behind it. These sophisticated systems learn the same way as your gesture recognizer, just with vastly more data and computing power.

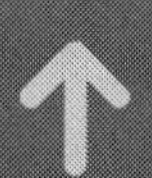

## MOVING FORWARD

The no-code AI movement is just getting started. What we see today is only the beginning, as these tools continue to evolve, making AI even more accessible and powerful for non-technical users. Barriers will keep lowering, capabilities will expand, and creating AI-driven solutions will become as common as building a website.

For beginners, no-code platforms offer the perfect starting point. They let you apply AI to real-world problems without the steep learning curve of traditional programming. They help you develop a deeper understanding of AI concepts that apply across the field.

As you experiment, you'll gain a practical sense of what AI can and cannot do. This growing "AI literacy" is valuable no matter what industry you're in. Even if you eventually decide to learn to code or collaborate with technical teams, the experience you gain through no-code platforms will give you a sound foundation.

In the next chapter, we'll dive into AI tools you can use today, from image and speech recognition to data analysis and content creation. Some build directly on the no-code principles we've covered, while others introduce new ways to harness AI without requiring technical expertise.

Today's technological revolutions don't belong only to those who can code. They belong to anyone willing to explore, experiment, and imagine new possibilities. No-code AI puts these powerful tools in your hands. What will you create with them?

## References

1. *Burt, C. (2019, May 1). Universities use AI to boost student graduation rates. EdTech Magazine. https://edtechmagazine.com/higher/article/2019/05/universities-use-ai-boost-student-graduation-rates*

2. *Blue Yonder. (2021). Migros reduces food waste and optimizes inventory with AI-driven demand forecasting. Blue Yonder. https://blueyonder.com/migros-ai-food-waste*

3. *Google. (2020, January 28). AI's killer (whale) app. The Keyword. https://blog.google/technology/ai/protecting-orcas/*

**Today's technological revolutions don't belong only to those who can code. They belong to anyone willing to explore, experiment, and imagine new possibilities.**

**"The real danger is not that machines will begin to think like humans, but that humans will begin to think like machines."**

**Sydney J. Harris**

CHAPTER 04

# Practical AI Tools You Can Use Today

Remember, getting your first smartphone? Not the sleek device you're probably carrying now, but that first clunky attempt at putting a computer in your pocket. I sure do. Mine was a Palm Treo, this brick-like contraption that seemed miraculous for the times. It could check email, browse a simplified version of the web, even play simple games.

What I couldn't do was talk to it. I couldn't ask it to analyze data for me. Definitely I couldn't have it generate images or write paragraphs or automate my boring tasks.

But today? The AI tools available to average people, not researchers or corporations with million-dollar budgets, would have seemed like science fiction just a few years ago. And the best part? You can use most of them today without writing a single line of code or understanding the complex math happening behind the scenes.

Let's explore the practical AI tools that are ready for you to use right now.

## Image and Speech Recognition Tools

I still remember the first time I saw advanced image recognition in action. It was around 2015, and I was helping a small retail business implement inventory management software. The owner casually mentioned she wished they could just take pictures of products rather than typing in details.

*"Actually, we might do exactly that."*

Using a rudimentary commercial image recognition API, we built a simple system that could identify basic product categories from photos. It worked maybe 70% of the time, which wasn't great, but it was a start. The owner found it astonishing, even with those limitations.

If she felt impressed then, today's possibilities would completely blow her away.

## What AI Can Do and Its Limits

AI today can accurately recognize objects, people, text, and activities in photos. It can detect emotions on faces, extract text from documents and signs, organize images into groups, and even describe entire scenes.

Speech recognition has also improved, reaching over 95% accuracy in the right conditions.1 Modern systems can distinguish between different voices, understand multiple languages and accents, function in noisy environments, and interpret context better than before.

These advancements have led to many new applications. Today, we have:

- Accessibility tools for visually impaired users that can describe the world around them.
- Real-time translation of both text and spoken language.
- Automated surveillance for security (with all the ethical questions that raises).
- Medical image analysis that can sometimes outperform human radiologists.
- Content moderation at scale across social platforms.

*But let's be real about the limitations too.*

Image recognition still struggles with things it hasn't trained on. It has trouble with unusual objects, images from underrepresented cultures, and telling apart similar things (like different bird species). It also struggles with decisions that need cultural knowledge and can show bias based on its training data.

Speech recognition has its own challenges. It may struggle with strong accents or dialects it hasn't learned well. Background noise can make it harder to understand speech, and it often gets technical terms or uncommon words wrong. It can recognize words, but understanding meaning, emotions, or sarcasm is still a challenge.

## Consumer Tools

The exciting part is how accessible these tools have become. Google Lens lets you identify objects, plants, animals, and text just by pointing your phone camera. It's built into most Android phones and is available as part of the Google app on iPhones. (We'll be working more with this application in the "Try It Yourself" at the end of this chapter.)

Microsoft created Seeing AI, a free app designed for the visually impaired that narrates the world around you, reading text, describing scenes, identifying products, and recognizing people.

Those fun face filters on Snapchat and Instagram use advanced facial recognition AI. Photo apps like iPhoto and Google Photos also use AI to sort your pictures by recognizing faces, objects, activities, and even locations.

There are many ways to use speech recognition. Voice assistants like Siri, Google Assistant, and Alexa are the most common examples. AI also helps with dictation software for professionals, transcription tools like Otter.ai and Trint that turn speech into text, Google Translate for real-time translations, and apps like Speechify read text aloud in a natural-sounding voice.

A friend of mine, a freelance writer with repetitive strain injury, completely transformed her workflow using dictation software.

*"I was skeptical at first," she said, "But now I can 'write' for hours without pain, and the technology has gotten good enough that I barely need to make corrections."*

Her experience shows something important: These tools aren't just cool tech. They're helping real people solve actual problems.

# AI for Data Analysis

Back when I managed IT for a manufacturing company, data analysis meant working with Excel spreadsheets, pivot tables, and sometimes writing custom database queries. It was useful, but it took a lot of technical skill and time.

Today, AI has democratized data analysis in ways that would have seemed miraculous back then.

## Visualization and Reporting Tools

Modern AI-driven data tools make it easier than ever to spot patterns and trends in large datasets. Instead of spending hours analyzing data, AI can quickly suggest the best charts, graphs, or reports based on the data type. Many of these tools even explain insights in plain language, build interactive dashboards without coding, and answer questions about your data.

For example:

- Microsoft Power BI suggests visualizations, detects trends, and lets users type questions like "Show me sales by region for the last quarter" to generate charts instantly.
- Google Data Studio connects with Google's machine learning tools for deeper analysis, even though it's not as AI-heavy.
- Tableau now includes natural language processing, so users can ask questions through its Ask Data feature.

Thanks to these tools, even businesses without data science teams can make smart, data-driven decisions.

## Predictive Analytics for Everyone

AI has made it simple for people without technical skills to use predictive analytics. With tools like Obviously AI, you can upload a spreadsheet, choose what you want to predict, and create a model without needing to write any code.

For example, a retail client of mine used Obviously AI to predict which customers were likely to buy again within 30 days. They simply uploaded their customer data, set "will purchase" as the target, and the system created a model with 83% accuracy. By focusing marketing efforts on high-value customers with a lower purchase probability, they increased repeat business by 23%.

In the background, the system takes care of complex tasks like organizing data, choosing important details, training the model, and checking accuracy. But the design is simple, so users can focus on business decisions instead of the technical stuff.

Of course, this ease of use comes with tradeoffs. These tools can't match the precision of a skilled data scientist working with custom models. But they deliver 80% of the value with just 20% of the effort, making predictive analytics accessible to small businesses and individuals who wouldn't otherwise have access to this technology.

## Content Creation with AI

Creating text, images, or videos used to take years of practice and special skills. Now, AI tools can do it for you, making it easier than ever to create high-quality content.

### Text Generation and Enhancement

AI writing tools have become much more powerful and easy to use. ChatGPT and Claude can write essays, stories, poems, ads, and more from simple prompts. Jasper focuses on marketing content, helping with ads, emails, and blog posts. Grammarly not only checks grammar, but also suggests ways to improve style and tone.

As part of research for this book, I interviewed a director working at a small marketing agency. They completely transformed their workflow using these tools. Rather than staring at blank pages, their writers started with AI-generated drafts based on outlines, then edited and personalized the content.

*"We're producing three times the content with the same team," the director said. "But more importantly, our creative people are spending their time on the high-value aspects of writing, not struggling with first drafts."*

## Image and Design Tools

AI has made tremendous advances in image creation. Tools like ChatGPT, Midjourney, Leonardo AI, and Stable Diffusion can turn simple text descriptions into stunning images. Canva's Magic Studio makes design easier by combining AI-generated images with ready-made templates. RunwayML helps users edit images and videos with AI, and Adobe has added AI-powered tools to its creative software with Firefly.

These tools are impressive. If you haven't tried them, I recommend checking out free options like ChatGPT or Leonardo AI to see what they can do. They make professional design easier for beginners and speed up creative work for professionals. And they keep getting better, with new features and updates arriving every month.

There are still some challenges. AI-generated images can have biases from their training data, have trouble with details like realistic hands, and raise questions about copyright and the role of human artists. But being able to create high-quality images with just a few words is giving everyone new creative opportunities.

## Audio and Video Creation

AI is changing how we create audio and video. Descript lets you edit audio by editing text and can even copy a speaker's voice. ElevenLabs creates realistic AI voices that can read any text out loud. Synthesia makes videos with AI avatars that speak your script, and Podcastle helps record and edit podcasts using AI.

AI video tools are improving quickly. They can turn text into talking head videos, place people into scenes they were never in, or even create full videos from descriptions. These tools bring new creative possibilities, but they also come with risks, such as deepfakes and misinformation.

While these AI tools are powerful, they can also be tricky to use. People should approach them with both excitement and caution. AI-generated content is fast and impressive, but it still needs careful human review. Images might show unexpected biases, text might include mistakes, and creative outputs need a thoughtful check.

The best way to use these tools is to see them as helpful assistants, not perfect solutions. They are great for making first drafts, sparking ideas, and handling repetitive tasks. However, humans still need to make the final decisions, consider the ethics, and add the creative touch.

## ? ETHICAL SPOTLIGHT ?

### Digital Manipulation and Authenticity

The content creation tools we have explored bring up important questions about trust. When AI can create realistic images, text, and voices that were never real, how can we trust what we see online? These tools offer exciting creative possibilities, but they also make it easier to spread false information.

Using AI responsibly means being clear when content is AI-generated. It also means thinking about how others might misunderstand or misuse what you create.

## Automation Tools

Perhaps the most practical application of AI for many people is automation, using intelligent systems to handle repetitive tasks that would otherwise consume valuable time.

### Email and Communication Automation

Email remains the bane of professional existence, with many specialized AI tools helping tame the inbox. As we learned in the "Try It Yourself" in chapter 2, the good news is many of the most popular email programs like Microsoft Outlook and Gmail have incorporated AI powered features as part of the product.

AI is improving more than just email. Calendly uses AI to find the best meeting times without endless emails. Fireflies.ai records, transcribes, and summarizes meetings automatically. Krisp removes background noise from calls and meetings for clearer sound.

A remote team I worked with completely transformed their meeting culture using Fireflies.ai.

> *"We realized most of our meetings could be half as long," the manager said. "With AI summaries capturing actions and key points, we didn't need to worry about missing something, so conversations became more focused."*

### AI for Task Management

AI helps people manage tasks more easily by keeping work organized, improving schedules, and reducing effort. Modern AI tools can check workloads, set priorities, and even notice problems before they happen. AI assistants can also break large projects into smaller steps, assign tasks, and summarize meetings to keep everything on track.

Some tools, like Asana's Work Assistant and ClickUp AI, go even further by automating status updates and suggesting better workflows. This makes project management smoother and more efficient.

The major benefit of these tools is that they take the stress out of managing tasks. Using AI for task management can save 5-10 hours a week, not just by saving time but also by reducing the mental effort of constantly deciding what to do next.

## Customer Service Automation

AI is making customer service easier by using chatbots and other tools to handle common questions. We'll explore this more in the next chapter, but here are some popular options:

- Intercom's Resolution Bot automatically answers common customer questions.
- Zendesk AI suggests replies based on past conversations.
- Ada lets businesses create AI chatbots with no need to code.

AI chatbots can significantly improve efficiency for small e-commerce businesses. A report by Juniper Research (2023) found that AI-powered customer service solutions handle about 35-45% of routine inquiries without human help. For businesses with fewer than ten employees, this automation saves an average of 15 hours of staff time each week. This allows employees to focus on growth activities instead of repetitive customer service tasks.2

As discussed earlier in the chapter, let's wrap this up with a hands-on dive into how AI "sees" the world.

### Image Recognition Exploration

This simple exercise will give you firsthand experience with computer vision technology, the same tech that's revolutionizing everything from healthcare to retail to transportation.

**Tools Needed**

Smartphone with camera and Google Lens (or similar image recognition app)

**Getting Google Lens (it's a free app)**

- Android users: Google Lens is likely pre-installed on your device. It's available in the Google app, Google Photos, or as a standalone app in the Play Store.
- iPhone users: Download the Google app from the App Store, then look for the Google Lens icon (a small camera icon in the search bar).
- Alternative: You can also use the Google Photos app on any device. Open a photo, then tap the Lens icon.

**Step 1: Gather Your Test Images**

Take 5 photos of different objects or scenes around your home or workplace. For the most interesting results, include:

- An everyday object (like a household item)
- Something with text (a book cover, product label, or document)
- A plant or food item
- A complex scene with multiple objects
- Something unusual or potentially challenging for AI to identify

**Step 2: Analyze Your Images**

For each photo:

1. Open Google Lens (either directly or through Google Photos)
2. Select or take the photo you want to analyze
3. Wait a few seconds for the AI to process the image
4. Note what the AI identifies, suggests, or searches for
5. Explore the different options Google Lens provides (shopping, translation, search, etc.)

**Step 3: Record Your Observations**

Create a simple table like this for your findings:

| Image | What AI Identified Correctly | What AI Got Wrong | Surprising Interpretations |
|---|---|---|---|
| 1. | | | |
| 2. | | | |
| 3. | | | |
| 4. | | | |
| 5. | | | |

**Step 4: Reflect on the Experience**

Consider these questions:

- What types of objects or features did the AI recognize most accurately?
- What seemed to confuse the system?
- Did lighting, angles, or background affect the AI's performance?
- How might the AI's performance impact its usefulness in professional contexts?
- Did you discover any capabilities you didn't expect?

This simple activity ties directly to many of the AI applications we covered in this chapter. The same technology you just used helps:

- Doctors find issues in medical images
- Farmers detect plant diseases from photos
- Developers build tools for the visually impaired

- Self-driving cars recognize roads and traffic signs
- Stores create checkout-free shopping experiences

The strengths and weaknesses you noticed reflect real challenges businesses face when using AI. When image recognition works well, it boosts efficiency and accuracy. When it struggles, it shows why human oversight is still important.

By trying Google Lens, you saw firsthand how AI is great at spotting patterns but still needs human judgment to make sense of results. Knowing both the power and the limit of AI is key to using it effectively in your daily life and work.

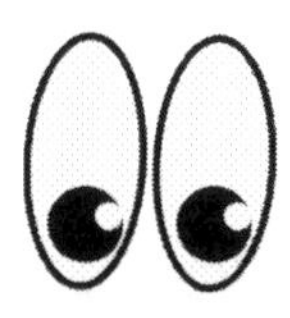

Next time you're hiking or gardening, try using Google Lens on a plant you don't recognize. The plant ID feature has helped me avoid poison ivy more than once!

## MOVING FORWARD

The tools we covered in this chapter are just a small sample of what's available, and AI is evolving fast. New features appear almost every week. What seems advanced today might be standard in a year, and entirely new AI tools will probably emerge.

You don't need to master every tool. The best approach is to experiment, find what works for you, and stay open to new options as they improve.

In the next chapter, we'll look at how businesses and entrepreneurs can use AI. You'll see how AI improves customer service, strengthens marketing, and makes operations more efficient. Many companies, big and small, are using these tools to grow without spending a fortune on technology.

Remember, the goal isn't to use AI for everything, but to identify the specific areas where it truly adds value to your life and work. The best technology is the kind that fades into the background, quietly making things better without calling attention to itself. That's the promise of practical AI, and it's available to you right now.

## When AI Tools Don't Cooperate

As you experiment with AI tools, you'll occasionally face situations where they don't perform as expected. On the next page, here's a quick troubleshooting guide to help you overcome common challenges:

Remember that even the best AI tools have limitations. When troubleshooting doesn't work, consider whether a different tool or approach might better fit your needs. Sometimes, the most effective solution is to recognize when AI isn't the right tool for a particular task.

### References

1. *Xiong, W., Wu, L., Alleva, F., Droppo, J., Huang, X., & Stolcke, A. (2018). The Microsoft 2018 conversational speech recognition system. 2018 IEEE International Conference on Acoustics, Speech and Signal Processing (ICASSP), 5934-5938. https://doi.org/10.1109/ICASSP.2018.8461994*

2. *Juniper Research. (2023). Chatbots: Critical role in e-commerce efficiency. Juniper Research. Retrieved from https://www.juniperresearch.com*

## When AI Tools Don't Cooperate

| Problem | Quick fix | Example |
|---|---|---|
| Vague outputs | Be more specific in your instructions | Instead of "Write about dogs," try "Write a 300-word explanation about border collie training techniques for first-time owners" |
| Incorrect information | Fact-check important details and specify reliable sources | "Summarize recent climate research from peer-reviewed journals published since 2020" |
| Irrelevant responses | Break complex requests into smaller steps | First ask for relevant categories, then request details about each one separately |
| Biased results | Ask for balanced perspectives or multiple viewpoints | "Show pros and cons from different perspectives" or "Provide three different approaches to solving this problem" |
| Tool seems stuck in a loop | Start fresh with a clear, simple request rather than repeatedly correcting | Begin a new conversation with refined instructions instead of multiple corrections |

# A Quick Note from the Author

I HOPE YOU'RE ENJOYING THE BOOK so far and finding it as inspiring and useful as I hoped it would be when I wrote it. Your time and attention mean so much to me, and I truly appreciate you choosing to dive into this journey with me.

If this book has sparked new ideas or helped you, I'd love to hear your thoughts (and results!). Sharing your feedback not only helps me grow as an author, but also helps others discover the book and join the community.

It only takes a moment to leave a review—just 60 seconds—and it makes a world of difference. Plus, I genuinely enjoy reading about your experiences and the unique takeaways you've found in these pages.

Thank you for your support and let's keep learning together!

To leave feedback, please return to where you purchased this book. For Amazon, please visit your Amazon Orders page or:

1. Open your camera app
2. Point your mobile device at the QR code below
3. The review page will appear in your browser

*Thank you!*

"The real voyage of discovery consists not in seeking new landscapes, but in having new eyes."

Marcel Proust

CHAPTER 05

# AI for Business and Entrepreneurs

I INTERVIEWED THE OWNER OF A LOCAL BAKERY for this book. When we first connected, she looked exhausted. Between managing inventory, scheduling staff, analyzing sales patterns, handling marketing, and occasionally stepping in to frost cupcakes herself, she was working 70+ hour weeks. When I suggested AI might help, she laughed. "That's for tech giants and corporations with massive budgets," she said. "Not for small businesses like mine."

Six months later, she agreed to a follow-up interview. Her voice was different. Energized. "I haven't worked a weekend in two months," she told me. "And profits are up 22%."

What changed? She started using affordable AI tools to automate her most time-consuming tasks. Nothing complicated or custom-built—just ready-to-use solutions that gave her the same efficiency benefits that big companies with large budgets and IT teams once had.

This is the power of AI today. Tools that were once too expensive or out of reach are now available to businesses of all sizes, from solo entrepreneurs to growing companies. And they're reshaping competition in major ways.

This chapter covers affordable AI tools for small and medium businesses. You'll find practical solutions to use right away. These tools can help you serve customers better, market more effectively, run your business more simply, and stand up to bigger competitors.

## Customer Service Applications

Customer service is one of the biggest ways AI is changing businesses. It allows companies to offer fast, reliable support to more customers with no need to hire a large team. This is reshaping how businesses of all sizes handle customer interactions.

### AI Chatbots and Virtual Assistants

Most businesses use chatbots and virtual assistants for customer service. Simple ones follow basic rules. Better ones can answer hard questions, find your customer info, and even tell how customers feel from their messages.

Modern AI chatbots typically fall into three categories:

1. **Rule-based bots** work like interactive FAQs, following set paths based on specific inputs. They are simple but effective for common questions.
2. **Intent-based bots** can figure out what customers mean, even when they ask questions in different ways. They sort questions and send them to either computer answers or real people when needed.
3. **Conversational AI** represents the most advanced category, capable of maintaining context throughout interactions, handling multiple questions in a single message, and adapting to conversational nuances.

Regardless of what type of chatbot you use, they highlight a critical insight about business AI. Success rarely comes from replacing humans entirely, but from strategically automating routine tasks so your human talent can focus on where they add the most value.

## Implementation Strategies for Small Businesses

You don't need big company resources to use customer service AI. Small businesses can easily use a few simple approaches:

Platform-based tools like Intercom, Zendesk, and Freshdesk include AI for customer service. They work well with your customer data and current tools.

Chatbot builders like Ada, ManyChat, Chatfuel, and Tars help you make chatbots for different channels without coding. You pay monthly based on how many messages you send.

For businesses with some technical capabilities, API-based solutions from companies like Rasa or Botpress offer greater customization potential while remaining accessible to non-developers through visual interfaces.

For this book, I interviewed a small vet clinic owner in Colorado. He started simple with a basic chatbot that scheduled appointments and answered pet care questions. This alone cut phone calls by 35% and let them stay open longer without hiring more staff. Later, they upgraded to a smarter system that could tell the difference between regular questions and urgent medical issues.

The key lesson? Start simple. Focus first on automating your most frequent customer interactions. Track what questions actually come up repeatedly rather than what you think customers ask. Build from there.

A surprising finding is that businesses often think customer service AI will save them money, but the real benefit comes from better customer experiences. Customers like 24/7 service, consistent answers, and fast responses, which build loyalty. HubSpot's 2022 report found almost 90% of customer service leaders say customer expectations are higher than ever, and

90% of customers want immediate responses. With 60% of customers wanting answers within 10 minutes, automation helps meet these expectations without hiring lots of staff.1

## Marketing and Sales Enhancement

AI is changing how businesses handle marketing and sales by helping them find potential customers, personalize messages, and improve campaigns, tasks that were once only possible for big companies with specialized teams.

### Predictive Analytics for Customer Targeting

Traditional marketing often relies on guesswork. AI-powered predictive analytics help by spotting patterns in customer data to predict what they will want or do next.

Even small businesses now have access to tools that can:

- Identify which prospects are most likely to convert
- Predict which customers might plan to leave (churn prediction)
- Determine optimal timing for outreach
- Recognize cross-selling or upselling opportunities
- Calculate customer lifetime value to optimize acquisition costs

In the past, only experts could use tools that predict business trends. Now anyone can use simple platforms like Obviously AI, Akkio, and PeopleDataLabs. You just upload your customer data and get useful predictions without writing any code. Small and medium businesses can now use these tools even without hiring data experts.

Here's something you might not know: you already have the data you need for good predictions. Your sales records, website numbers, email stats, and customer lists hold patterns that can help your business. AI tools find these hidden insights for you. You don't need to become a data expert to use them.

### Content Personalization and Optimization

Beyond identifying who to target, AI excels at determining what content will resonate with each audience segment and how to optimize its delivery.

Content recommendation engines, like those used by streaming services, help businesses suggest products or content based on what users do and like. Dynamic website personalization changes page content based on who's visiting (new or returning visitors, where they came from, what they did before). Email marketing tools use AI to find the best times to send emails, write better subject lines, and create content for different groups.

For content creation, tools like Jasper, Copy.ai, and Phrasee help businesses write marketing copy that matches their brand voice and gets more people to buy. These tools learn from your existing content and results to suggest messages that will connect with specific customer groups.

Research by McKinsey shows that personalization can boost revenue by 5-15% and make marketing money work 10-30% better.2 Big companies used to be the only ones with these advantages, but now AI tools make them available to all businesses.

## Operations and Efficiency

AI isn't just helping with customer service. It's also making companies work better on the inside. AI tools help businesses get more done with fewer resources by taking over boring tasks, using supplies better, and improving how products move from factory to store.

### Workflow Automation

AI helps find and fix repeated tasks that people used to do by hand. There are many ways businesses use this. For example, AI can sort through paperwork, process orders, and handle basic emails without human help. This frees up workers to focus on more important jobs.

- Document processing systems that can extract, classify, and route information from invoices, forms, and correspondence.
- Order processing automation that handles routine orders without human intervention.
- Meeting scheduling tools that eliminate the back-and-forth of finding available times.
- Intelligent email management categorizes, prioritizes, and sometimes responds to messages automatically.

Through my research for this book, I noticed a key pattern:

*When implemented thoughtfully, operational AI doesn't just cut costs—it enhances both customer and employee experiences by removing friction from everyday interactions.*

### Resource Optimization

Beyond automating workflows, AI excels at optimizing how businesses allocate their resources, from staff scheduling to inventory management.

AI tools create better staff schedules by studying past sales, employee wishes, and customer shopping habits. These systems also help businesses use their equipment and space more wisely. For energy costs, AI can predict when you'll use more power and adjust systems to save money.

I once helped a retail chain use AI to schedule their staff. The system looked at three years of sales and found some surprises. Everyone thought Tuesday mornings were slow, but the data showed many big spenders shopped then. It also found each store had its own busy times on Saturdays.

With these smarter schedules, sales grew by 18% per working hour. Staff were happier too because they got more of the shifts they wanted. This wasn't just about cutting hours. It was about having the right people working when and where customers needed them.

Businesses using AI to manage resources can save money while serving customers better. Some have cut energy costs by 35%.3 For small businesses with tight budgets, these savings can mean staying open rather than closing down.

### Supply Chain Improvements

Small businesses struggle with supply chains. They can't demand better terms like big companies can, but they still need to manage inventory, shipping, and orders. AI now offers simple tools to help:

- Inventory optimization tools that predict demand patterns and recommend optimal stock levels.
- Order forecasting systems that anticipate customer needs based on historical patterns and current signals.
- Supplier evaluation platforms that analyze performance data to identify risks and opportunities.
- Logistics optimization that determines the most efficient routing and fulfillment strategies.

Small businesses can easily start using AI for their supply chains. You already have the data in your ordering and sales records. Even slight improvements in predicting what you'll need can save you a lot of money.

## AI for Solopreneurs and Small Teams

Today's AI tools help one-person businesses and small teams do work that once needed many people. Tools exist that allow anyone to create content, serve customers, and study markets.

### Growing Without Hiring More People

Small businesses used to grow by hiring more people, which costs a lot of money. AI gives another option. You can serve more customers while keeping your team small.

Tools like Otter.ai can write meeting notes and track tasks. Clara or Calendly can schedule meetings for you. FigureIt can help with financial planning. These do jobs you might otherwise hire someone to handle. With tools like Canva for design, Descript for audio and video, and Jasper for writing, small teams can create professional materials quickly.

I talked with a business consultant who used AI tools to grow her practice. She went from 8 clients to 22 without hiring another employee. She used AI for scheduling, taking notes, and following up with clients. This saved her hours of paperwork. AI research tools helped her prepare for meetings twice as fast. She also used AI to create custom reports and proposals using templates.

> *"I used to spend 60% of my time on tasks that weren't directly serving clients," she said. "Now that's down to about 20%, which means I can take on more clients while actually delivering better service."*

You don't need many AI tools. Just find what slows down your business and fix those problems. Look for tasks that take too much time but add little value. These are perfect for using AI.

## ? ETHICAL SPOTLIGHT ?

### Automation and Workforce Impacts

When businesses use AI automation tools, they face important questions about how it affects employees. While AI often helps workers rather than replacing them, it still changes jobs and may remove some positions. Using AI responsibly means being open with staff, offering training for new skills, and thinking about the wider social effects of automation, not just the business benefits.

## Competing with Larger Organizations

AI also helps small businesses offer services that only big companies could provide before. This makes things more equal in several respects.

Small businesses can now create personal experiences for customers without hiring data experts. AI helps turn your business information into useful insights without needing special analysts. You can manage posts on many platforms when AI helps create and share content. You can offer excellent customer service by using AI to handle simple questions.

Now that we've explored how AI can enhance business operations, let's put these tools to the test in a hands-on challenge that will help you compare and select the best AI assistant for your needs.

## AI Content Analysis Challenge

Let's explore how different AI writing systems approach the same task, revealing their unique capabilities, limitations, and "personalities" which are crucial insights for selecting the right AI tools for your specific business or personal needs.

### Tools Needed

Access to three different AI writing assistants (ChatGPT, Claude, Gemini, Bing AI, or others available to you)

Most AI assistants require you to create a free account using an email address or Google account.Here are links to popular options (with free tiers):

- ChatGPT: https://chat.openai.com
- Claude: https://claude.ai
- Gemini: https://gemini.google.com
- Bing AI: https://www.bing.com/chat

We talked earlier about how AI tools are good at different things for businesses. Let's compare them side by side so you can see which one works best for you.

### Step 1: Pick What You Want Written

Choose one writing task for different AI tools to handle. Pick something you care about for work or your personal life. Here are some ideas:

- Product description for something you use regularly
- Brief explanation of a concept in your field for beginners
- Professional email requesting information or a meeting
- Creative piece like a short poem or story opening

The goal is to choose something concrete enough to compare results but open, ended enough to reveal differences in approach.

**Step 2: Create Your Prompt**

Write a clear, specific prompt that you'll use with each AI system. Here are example prompts for each suggestion from Step 1:

**For a product description:** "Write a 100-word product description for a waterproof Bluetooth speaker that would appeal to outdoor enthusiasts. Highlight key features and benefits."

**To explain a concept:** "Write a 150-word explanation of how artificial intelligence works for a 12-year-old with no technical background. Use simple analogies and avoid jargon."

**For a professional email:** "Write a professional email to a potential client requesting a meeting to discuss my graphic design services. Keep it under 200 words, be friendly but professional, and include a clear call to action."

**For a creative piece:** "Write the opening paragraph of a mystery story set in a small coastal town. Create an intriguing hook that introduces the main character and hints at a mysterious event."

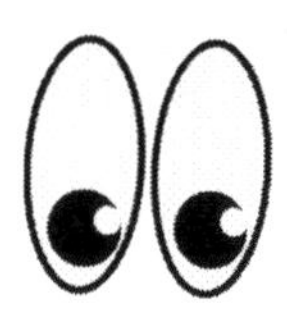

Save your chosen prompt somewhere you can easily copy/paste it to ensure you're giving each AI the same instructions.

**Step 3: Gather Responses**

Visit each AI writing assistant and submit your identical prompt to each one. If possible, use them in their default modes without special instructions about being an expert or using a specific style.

1. Open the first AI tool (e.g., ChatGPT)
2. Paste your prompt and submit it
3. Save the response (copy to a document where you can compare all results)
4. Repeat with each AI tool

**Step 4: Compare and Analyze**

Create a simple comparison chart like this:

| Feature | AI Tool #1 | AI Tool #2 | AI Tool #3 |
|---|---|---|---|
| Word Count | | | |
| Tone and Style | | | |
| Specific details Included | | | |
| Unique approaches | | | |
| Structure | | | |
| Strengths | | | |
| Weaknesses | | | |

As you fill in this chart, consider:

- Which AI created content that best matched what you were looking for?
- Did any AI misunderstand aspects of your request?
- Were there notable differences in creativity, formality, or structure?
- Did any include unexpected elements (like disclaimers, questions, or suggestions)?

**Step 5: Test Adaptability**

If you have time, try a follow-up experiment. Take your original prompt and add specific instructions like:

- "Make it humorous and casual"
- "Use technical language appropriate for experts"
- "Include three bullet points highlighting key benefits"

See how each AI adapts to this additional requirement. Does one handle the change better than others?

## CONNECTING TO THE CHAPTER

This exercise directly shows several key concepts from this chapter.

1. **Customer Service Applications:** You've experienced how AI can generate customer-facing content, connecting to our discussion of chatbots and virtual assistants.
2. **Marketing Enhancement:** You've seen how AI can create marketing materials and content personalization, as covered in our marketing section.
3. **Operational Efficiency:** You've explored how AI can streamline business communications and workflows, relating to our discussion of workflow automation.
4. **Small Business Scaling**: You've practiced using AI tools that allow small teams to produce professional-quality business content efficiently.

This hands-on experience should help you better understand why we emphasize not just knowing these tools exist, but understanding their distinct capabilities when applying them to real-world business challenges.

## MOVING FORWARD

In this chapter, we explored how businesses of any size can use AI to support customers, improve marketing, boost efficiency, and stay ahead of competitors. These tools are available now, and you don't need tech expertise or a big budget to use them.

Business is changing quickly. According to Accenture, 84% of business leaders believe AI is essential for growth.4 Still, many small and mid-sized companies think AI is only for the future or for large corporations. This creates an opportunity for those who start now.

Using AI doesn't mean replacing people. It means handling routine tasks with AI so your team can focus on creativity, empathy, and strategy. The goal is to combine AI and human skills, not choose one over the other.

Next, we'll look at how AI is shaping specific industries like education, healthcare, creative work, and security. You'll see real examples and get ideas you can apply in your own business.

## References

1. *HubSpot. (2022). Annual state of service report. HubSpot Research. https://www.hubspot.com/hubfs/assets/flywheel%20-campaigns/HubSpot%20Annual%20State%20of%20Service%20Report%20-%202022.pdf*

2. *Boudet, J., Gregg, B., Rathje, K., Stein, E., & Vollhardt, K. (2019). The future of personalization—and how to get ready for it. McKinsey Quarterly.*

3. *Rahd AI. (2023, December 5). AI tech could save billions in North Sea decommissioning. The Times. https://www.thetimes.co.uk*

4. *Accenture. (2022). Technology Vision 2022: Meet Me in the Metaverse. Accenture Research.*

"Industry leaders don't offer different tools; they use familiar tools differently."

Morgan Housel

CHAPTER 06

# AI Across Industries: Key Applications

BACK WHEN I FIRST STARTED CONSULTING on IT projects, we used to have this saying: "Technology doesn't solve problems; people with technology solve problems."

Those words still ring true today, especially with artificial intelligence. AI's superpower isn't about fancy tech that sounds impressive. It's about how people use these tools to solve problems in their specific line of work.

## A Guided Tour Through AI's Cross-Industry Impact

In this chapter, we will take a broad look at the use of AI in different industries. Instead of focusing deeply on one field, we will explore a variety of examples. This will give you a better understanding of the applications of AI's core abilities, such as recognizing patterns, making predictions, automating tasks, and personalizing experiences to solve various problems.

As we journey through these diverse sectors, look for these recurring themes:

1. **Problem-Specific Applications:** How each industry adapts similar AI technologies to address their unique challenges
2. **Human-AI Collaboration:** The consistent pattern of AI handling routine tasks while humans provide judgment and creativity
3. **Implementation Approaches:** How organizations start small, focus on specific problems, and gradually expand AI use

Depending on your interests, some sections may be more useful to you than others. If you work in education or healthcare, those parts will give you direct insights for your field. If you are an entrepreneur or business professional, seeing patterns across industries may help you find new ways to apply AI. Creative professionals may focus on the sections about content creation and design. And if you are simply curious about AI's impact on society, the wide range of examples will give you a clear picture of how these technologies are changing different fields.

This overview of industries won't make you an expert in each one, but it will broaden your understanding of what AI can do. It will help you notice patterns and uses you might not have considered. With this bigger picture, you will be better able to spot opportunities in your own field, even if not directly covered in this book.

Now, let's begin our exploration of how different industries are putting AI to work in practical, meaningful ways.

## Education and Learning

Our old education system served a different time. It created uniform workers for factories and industries. But the world has changed. Now we need workers who can think creatively and solve complex problems. AI is helping schools meet these new challenges by giving teachers powerful tools to personalize learning in ways they couldn't before.

## Personalized Learning Platforms

*"Every teacher knows we should personalize education, but with 30 kids in a classroom, how are you supposed to do that?"*

An eighth-grade math teacher asked me that question during an interview. For years, he struggled to create different lesson plans and tests for students with different learning needs. The paperwork and preparation were overwhelming. Now, everything has changed. With an AI tool called ALEKS (Assessment and LEarning in Knowledge Spaces) developed by McGraw Hill, his classroom runs differently. Students can learn at their own speed, following personalized learning paths. The system tracks what each student knows and automatically adjusts the content to match their skill level.

*"The system knows exactly which concepts each student has mastered and what they're ready to learn next," he said. "I can finally teach the way I always wanted to, focusing my attention where it's needed most instead of teaching to the middle and losing both ends of the spectrum."*

Scientific research backs up this approach. An extensive study by the RAND Corporation showed that students using personalized learning tools learned more than students in traditional classrooms did. Students who were struggling before saw the biggest improvements.1

What makes these learning platforms work isn't just about delivering information. They're smart about tracking student progress. Tools like DreamBox Learning, Lexia Core5, and Carnegie Learning's MATHia do more than just show lessons. They watch how students work through material, spotting exactly where students might get confused or struggle to understand.

### Assessment and Feedback Tools

Grading and providing feedback has always been the most time-consuming part of teaching. Writing teachers can spend entire weekends grading essays. Science teachers might take hours checking lab reports. Teachers face a tough choice: either give fewer meaningful assignments or give up their personal time to provide helpful feedback.

AI is changing this problem. New tools can now give instant, consistent feedback, which frees teachers to focus on more important guidance.

In science and math classes, tools like Gradescope use AI to check student work and spot common mistakes. This helps teachers see when the entire class is having trouble with a specific topic. Now teachers can quickly address those challenges before moving on.

These tools aren't about replacing teachers. Instead, they help teachers work smarter. By handling routine grading tasks, AI frees teachers to do what they do best: inspire, guide, and connect with students.

### Accessibility Enhancements

The most exciting part of AI in education might be how it helps students with different learning needs. Tools that used to be special accommodations are now becoming helpful for everyone.

New technologies can now read text out loud and convert speech to writing. This means students with reading or visual challenges can access learning materials on their own. Other tools can break down complex texts into simpler language, helping struggling readers and students learning English understand grade-level content more easily.

I visited a rural high school implementing Microsoft's Immersive Reader across the curriculum. A student with dyslexia told me,

*"Before, I needed a special education teacher to read tests to me. It was embarrassing to leave class. Now I can use text-to-speech right on my laptop. No one even knows if I'm using it, and I can work at my pace."*

Students with hearing difficulties can now use apps that turn speech into text in real time. Tools like Otter.ai can tell different speakers apart and recognize complex words. Research shows that technology designed for students with disabilities helps everyone. For example, video captions improve understanding and memory for all students, not just those with hearing loss.

This approach means schools are building helpful tools right into their learning platforms instead of treating them as special add-ons. A student who doesn't have any diagnosed learning challenges might use text-to-speech to help understand a reading assignment better. Or they might use speech-to-text to capture their ideas quickly when writing feels slow or difficult.

ETHICAL SPOTLIGHT

### Fair Access to AI Benefits

When examining AI's use across industries, we should assess the fair distribution of its benefits. For example, advanced healthcare AI might mostly help wealthy hospitals, while poorer communities see little improvement. In education, AI could increase the gap between well-funded schools and those with fewer resources. Using AI responsibly means making sure its benefits reach people from all backgrounds, no matter their income or resources.

## Healthcare Innovations

Healthcare is where AI can truly save lives. By combining huge amounts of medical data, complex decision-making, and high-stakes situations, AI offers exciting ways to help doctors do their jobs better.

### Diagnostic Assistance

AI is becoming a valuable tool in medical diagnosis. It can analyze medical images, test results, and patient records to find patterns that even skilled doctors might overlook.

A study in Nature Medicine showed that an AI system screening for breast cancer caught more potential cases than radiologists working alone. The study showed a 9.4% reduction of false negatives and 5.7% reduction of false positives than the radiologist working alone.[2] This means thousands of patients could get more accurate diagnoses and treatment.

In labs, AI helps doctors check tissue samples for cancer. These tools work with human experts to find more possible cancer cases and make fewer mistakes. A study in JAMA Oncology found that AI helped pathologists be more accurate by reducing both missed cases and false alarms. The research showed that AI improved detection without lowering accuracy, leading to better cancer diagnoses.[3] This helps doctors find serious illnesses sooner and with more confidence.

Epic Systems created the Epic Sepsis Model (ESM), an AI tool that helps doctors spot sepsis early by checking patients' vital signs and lab results. Sepsis is a dangerous condition where early treatment can save lives. The ESM connects to electronic medical records and updates risk scores every 20 minutes. This gives medical staff a chance to act sooner than with traditional methods.

However, a study by the University of Michigan found that the tool's accuracy depends on the timing of the data collection. The model is less reliable when it only uses early patient data before sepsis symptoms appear. This shows the need for more testing to confirm how well it works in real-world hospitals.4

What makes these diagnostic tools powerful is their ability to keep getting better. Every new medical case helps the AI learn more. This means the system becomes more accurate over time and can handle more complicated medical diagnoses.

### Personalized Treatment

Doctors usually follow standard treatments that work for most people. AI is helping to change this by creating treatment plans that fit each patient's unique needs.

In cancer treatment, tools like IBM's Watson look through massive amounts of medical research and patient information. They help oncologists find the most promising treatment options for a specific patient. While these systems weren't perfect at first, they've gotten much better at sorting through complex medical information.

In mental health, some companies are developing smart tools that track changes in how people use their phones. By looking at things like typing speed and word choices, these systems can spot early signs of depression or anxiety. This means doctors might catch mental health challenges earlier than they could with regular appointments.

Another cool thing AI can do is help with figuring out how much medicine someone should take. You know how tricky medicines can be. Too little and they don't work, but too much and you might get sick from side effects. Well, these new computer programs can look at everything about a person and suggest the exact right amount of medication for them. This is super helpful for medicines where getting the dose perfect is really important.

### Administrative Efficiency

While AI that can diagnose diseases gets all the buzz, some of the most useful AI in healthcare is actually tackling the mundane stuff: paperwork. These AI tools are like super-efficient administrative assistants. They're saving hospitals money, helping more people get the care they need, and most importantly, freeing up doctors and nurses to spend more quality time with patients instead of being buried in forms and files all day. It may not sound as flashy as AI playing doctor, but to make a real difference in people's lives, these paperwork-busting tools are the unsung heroes of healthcare AI.

AI-powered tools like Dragon Medical One are changing how doctors handle medical records. At Temple Health, doctors saved over 3,600 hours each month by using this system. One doctor discovered he saved up to five minutes per patient, which adds up to more than a month of work time saved each year. This means less stress for doctors and more time to care for patients.5

I saw this work firsthand at a regional hospital. Before implementing AI scheduling, the hospital experienced an 18% patient no-show rate, and some departments remained unused. The AI looked at past patterns and predicted which appointment times different patients were most likely to keep. The result? Their missed appointment rate dropped to 7%, and they used their resources more efficiently without hiring more staff.

Another area where AI helps is handling insurance and billing. Systems like Olive can automatically check insurance, get treatment approvals, and process claims. One healthcare system saved $4.2 million annually by using this technology.6

These might seem like boring improvements, but they matter a lot. By cutting down on paperwork, AI helps reduce doctor burnout. It makes healthcare more cost-effective and helps more people access care. Most importantly, it lets medical staff focus on the human parts of healthcare that machines can't do.

## Creative Industries

Artists and writers used to think creativity was something only humans could do. But AI has changed that thinking. Now, artificial intelligence can create artwork, music, and writing that surprises and challenges our old ideas about creativity. We touched on this earlier in the book, but here we're going deep here with the benefits for creatives.

## AI in Art, Music, and Writing

Generative AI systems have transformed the creative landscape by producing original content based on text prompts or learning from existing works.

Artists now have powerful new tools like ChatGPT, Midjourney, and Stable Diffusion that can create images just from written descriptions. These tools are changing how people make art. Now, someone who never took an art class can create illustrations, design concepts, and try out creative visual ideas.

In music, AI tools like AIVA can now compose entire musical pieces in different styles, from classical symphonies to modern pop. Musicians can use these tools to break through creative blocks, explore fresh sounds, or quickly generate background music for videos and other projects.

In writing, large language models like GPT-4 can generate text across a wide range of styles and formats. These tools help writers overcome blocks, explore different approaches to a topic, or handle routine writing tasks so they can focus on more creative or strategic aspects.

A novelist I interviewed uses AI as a brainstorming partner.

> *"When I'm stuck on plot development or character arcs, I prompt the AI with different scenarios to see how they might play out," she said. "It's like having a writing partner who never tires of exploring possibilities. I still make all the creative decisions, but it helps me consider options I might not have thought of otherwise."*

Creative professionals discover AI tools can speed up their work. In a recent study, researchers found that people using AI writing helpers finished tasks about 25% faster and produced better work.7 The AI handled the boring, repetitive parts of the job, which let workers focus on the more interesting and creative elements.

But not everyone is thrilled. Some workers worry about how AI might change their creative process. They wonder if machines will take over too much of their work or make their jobs feel less personal.

### Collaboration Between Humans and AI

The most interesting creative applications aren't about AI replacing human creativity but augmenting it through collaborative processes where each contributes different strengths.

An architect shared how her team uses AI in design. The computer helps them quickly explore thousands of design options for buildings. It looks at things like the use of space, how strong the structure might be, and how much energy it might need. The AI doesn't actually design the building. Instead, it shows possibilities that humans might miss. The computer handles the complex math, while the architects focus on making the design beautiful and making sure it works for people.

This approach is showing up in other creative fields too. In film and video production, AI can now do technical tasks like separating actors from backgrounds, adjusting colors, and even creating rough cuts of videos. This means smaller film companies can now do work that used to require huge teams.

The best part? This isn't about robots taking over creative jobs. It's about AI being a helpful tool that makes creativity easier. These tools help more people express themselves and give professionals new ways to do their work. Humans are still in charge, using AI like a super-smart assistant.

## Cybersecurity

These days, we do more and more of our work on computers and the internet. Because of this, keeping our computers and information safe from bad guys has become super important for businesses. It's not just a job for the IT department anymore. Now, the entire company needs to work together to secure everything from hackers and other threats.

### Threat Detection and Prevention

Artificial intelligence is changing how we protect computer systems. In the past, security systems only looked for attacks they already knew about. This meant new types of attacks could slip through unnoticed for months.

Now, AI-powered security tools work differently. They learn what normal computer behavior looks like. When something strange happens - like someone accessing files in the middle of the night - the system raises an alarm.

I saw this work firsthand at a manufacturing company. Their AI security system spotted unusual activity at 2:30 AM. The computer noticed something was off, even though it didn't look like any attack the system had seen before. It quickly blocked the suspicious activity, preventing a potential disaster that could have cost the company millions.

*Think of it like a smart guard that learns the normal patterns of a building. If someone tries to sneak in at an odd time or through an unusual door, the guard immediately notices and takes action.*

Research from the Ponemon Institute found that organizations using AI-powered security tools detected and contained breaches 27% faster than those using traditional approaches, reducing average breach costs by $1.1 million.8

Some computer security tools now use smart technology to learn what's normal for a specific company. They look for tiny signs of trouble that humans might miss. Unlike older systems that only watched for known viruses, these new tools can spot unusual behavior that might signal a new type of attack.

For example, Darktrace learns a company's typical computer behavior. Vectra AI looks for how attackers might try to break in. CrowdStrike's system watches for suspicious actions in real-time.

The advantage? These tools can handle massive amounts of information. A medium-sized company might create millions of computer security events every day. Humans can't possibly watch all of that. But AI can quickly scan through everything, pick out the most important warnings, and ignore false alarms that might make security teams tired and less alert.

It's like having a superhuman security guard who never gets tired and can watch every single door and window at the same time.

## Protecting Privacy

Companies collect more personal data than ever, raising concerns about safe data usage. AI offers innovative solutions to this challenge.

**Federated learning** is an approach which allows organizations to collaborate without sharing sensitive information. Instead of pooling all data in one place, the AI model travels to each organization's data separately. For example, hospitals can improve medical research without exposing patient records. The model learns from each hospital's data locally, then combines these insights without ever seeing the actual patient details. It's like solving a puzzle where each hospital keeps their piece private but shares what they learned from it.

Another method, **differential privacy**, adds carefully calculated "noise" to data. This makes identifying specific individuals nearly impossible while maintaining the data's usefulness for analysis. Apple uses this approach to improve products without tracking exactly what each user does.9

These approaches are gaining traction. A recent study found that 60% of companies now have AI ethics guidelines, with tools helping organizations map their data, assess privacy risks, manage permissions, and detect potential issues.10 The goal is balancing data innovation with protecting personal information.

### AI in Cybersecurity

Cybersecurity is like a constant battle between hackers and security teams, with both sides always finding new ways to attack or defend computer systems. Now, artificial intelligence is being used as a powerful tool by both hackers and defenders.

Attackers are getting smarter by using AI to create more convincing scam emails. They can now scan social media and public information to craft messages that look exactly like they're from your coworkers or boss. These fake emails are so detailed that they're hard to spot as fraud.

On the defense side, companies are using AI to test their own computer systems. Think of it like hiring a team of super-smart hackers to break into your own building - but without actually causing damage. These AI systems probe for weaknesses, checking both technical defenses and how easily employees might fall for tricks.

In the future, AI will play a bigger role in cybersecurity, with AI systems fighting against each other. Human experts will focus more on big-picture decisions and handling tricky threats that AI can't manage yet. This means security professionals need to learn new skills. Instead of just looking for known problems, they'll need to understand how AI works, what it can and can't do, and how to guide these smart systems.

## Agriculture and Environment

Farmers today face an immense challenge: growing enough food to feed everyone while protecting the environment. Artificial intelligence is helping them to do just that in new and creative ways.

## Precision Farming

In the past, farmers treated entire fields the same way. They would spread fertilizer or water across every acre, even if some areas needed more or less. Now, smart technology lets farmers take a more precise approach.

I saw this work on a farm in the Midwest United States. The farmers used sensors in the soil, weather stations, and satellite images to understand exactly what each part of their field needed. Their computer system could tell them:

- How much water each area requires
- Where to use fertilizer
- Which spots might have pest problems

The farm manager shared an amazing story with me. Before using AI, they spread the same amount of fertilizer across the entire farm. Now, they can provide each section with exactly what it needs. The results were impressive: 22% less fertilizer used, 7% more crops grown, and a boost to both their wallet and the planet.

Some cool AI farming tools can now spot individual weeds and spray only those spots.11 They can also detect crop diseases early and help farmers make smarter decisions.12

The best part? These tools are becoming cheaper and easier to use. Farmers can now use smartphones and simple sensors to get advanced farming help. This means even small farms in poorer countries can improve their crops.

## Climate Modeling and Conservation

Beyond farm-level applications, AI is transforming how we understand and respond to environmental challenges at regional and global scales.

Studying climate change used to need big, costly computers. Now, AI helps us understand it faster and more accurately. AI is becoming a powerful tool for protecting the environment. Here are some amazing ways it helps:

### Tracking Forests and Wildlife

- Some organizations use sound sensors to catch illegal logging
- AI can listen for chainsaws or vehicles in remote areas
- When it hears something suspicious, it immediately alerts authorities

### Satellite Watching

- AI can track tree cover from space
- It helps scientists follow animal migrations
- This technology can predict where wildfires might start
- It can help farmers use water more wisely

Why does this matter? Big environmental problems like climate change and animal extinction have always felt too hard to fix. AI helps us better understand these issues, find problems sooner, and come up with smarter solutions.

Think of AI as a super-smart detective for the planet. It can see patterns and connections that humans might miss. The technology works 24/7, covering enormous areas that would be impossible for people to monitor alone. By analyzing massive amounts of data and recognizing subtle changes, AI helps us approach environmental challenges with more insight and precision than ever before.

## Transportation and Logistics

AI systems now change how we move people and things. Just a few years ago, we couldn't imagine self-driving cars or smart maps, but now they exist. Computers help trucks find faster routes, make deliveries more efficient, and can even drive vehicles. This makes travel quicker, less expensive, and often safer.

## Autonomous Systems

Autonomous vehicles represent one of the most ambitious applications of AI, integrating computer vision, sensor fusion, decision-making systems, and control algorithms to navigate complex, unpredictable environments.

Driverless cars aren't quite ready to hit the streets, but they're already hard at work in some special places. Mining companies have big trucks that drive themselves all day and night. They don't need breaks! Warehouses use robotic forklifts and pallet movers that zip around on their own. And ports have self-driving vehicles that carry shipping containers from place to place.

I visited a warehouse where a fleet of robot helpers has completely changed how things work. The manager there said,

> *"The robots work side-by-side with our human team. They handle the boring, repetitive stuff like moving things around. Meanwhile, our people focus on more complicated tasks like picking orders and packing boxes. Since we brought in the robots, we're getting 37% more work done without needing a bigger building. And get this: Fewer workers are getting hurt from repeating the same movements over and over."*

Public transportation is slowly testing out driverless vehicles too, but they're starting in places that are easier to control. A few airports now have shuttles with no drivers that carry people between terminals. Cities like Singapore are trying out self-driving buses, but only on specific, pre-planned routes. This helps people get used to the idea of vehicles with no one behind the wheel, without the craziness of cars and pedestrians darting every which way.

## Route Optimization

AI is changing transportation by improving route planning. Old systems used fixed maps and past data, but modern ones adjust in real time, predict traffic, and keep routes running smoothly.

Have you ever wondered how delivery companies like UPS figure out the best way to get your packages to you? They use a special computer system. It looks at things like traffic, which packages need to be delivered first, and what each driver can do. This program helps UPS save a ton of gas and gets your stuff to you faster.

When you take an Uber or Lyft ride, there's also a smart computer in charge of figuring out which car should pick you up. It's always watching where people are and where they want to go. If many people suddenly need rides in one area, it'll send more drivers there. That's why you rarely have to wait too long to get a ride.

Even the buses and trains are using smart computers. In some cities, the computer looks at how many people are riding at different times and where they're going. If there are more people than usual, it can send more buses or make them come more often. This helps the buses and trains run better without needing more vehicles.

I talked to someone who plans how buses and trains run in their city. They said,

> *"Our computer looks at how many people rode in the past and how many are riding right now. It also pays attention to the weather and special events. If something unusual happens, like a big concert, it can change the schedule, so everything keeps running smoothly. It's especially helpful when big changes happen, like during the pandemic when suddenly fewer people were traveling."*

## Traffic Management

AI now changes how we control city traffic. Old systems used fixed light timers or basic sensors. New systems use cameras to see traffic, predict problems before they happen, and adjust signals as conditions change. These smart systems can respond quickly when traffic gets heavy or accidents occur, unlike the older fixed systems.

I worked on a project with a city government where they started using a smart computer system to control traffic. They put cameras at busy intersections that could see how many cars there were. The computer used this information to decide how long to make the traffic lights stay green or red. Before, the lights would just change based on a set schedule, even if there weren't many cars around.

In big cities, they have even fancier systems. These use information from lots of different places. They have cameras and sensors on the roads. Some cars can even "talk" to the computer and tell it where they are. The computer also looks at location information from people's smartphones (but don't worry, they remove any info that could identify a person). By looking at all of this, the computer can actually predict when traffic is going to get bad. It can then do things to help, like making the traffic lights change differently, putting slower speed limits on electronic signs, or sending messages to navigation apps to tell drivers to take a different route.

The city of Pittsburgh, Pennsylvania, USA, installed a smart traffic system, and it made a big difference. People's trips got 25% faster, and they spent 40% less time sitting still in traffic.13 The system is smart because it can make each intersection work better on its own, but it also makes sure they all work together. It can make a string of green lights so cars can keep moving smoothly down a whole road.

What's really great about these smart traffic systems is that they help cities handle more traffic without having to improve roads. It's hard and expensive to add more roads, especially in crowded cities. The computer systems make the roads we already have work better. They cut down on traffic jams, pollution, and wasted time.

TRY IT YOURSELF

## Identifying Business Automation Opportunities

Take a moment to mentally review your typical work week. As you consider your daily tasks and business processes, look for challenges that might be ripe for AI automation:

**Common Areas to Explore**

- Administrative tasks
- Communication management
- Content creation
- Customer follow-up
- Scheduling and time management
- Marketing and social media
- Reporting and data analysis

**Reflection Prompts**

For each task you identify, consider:

1. How time-consuming is this task?
2. Does it involve repetitive steps?
3. Are there obvious patterns or rules that guide the task?
4. Would a consistent, quick response be valuable?
5. Could an AI tool handle at least part of this process?

**Quick Assessment**

Ask yourself:

- Would automating this task save you significant time?
- Could AI handle the task with minimal human oversight?
- Are there potential risks in automating this process?

The goal is simply to notice where AI might offer potential efficiency gains in your work. Don't worry about finding a perfect solution; just begin to see your workflow through an AI-possibility lens.

## MOVING FORWARD

As we finish looking at the use of AI in different jobs, the most important thing to remember isn't about the computers or programs. It's about how people use AI to solve problems. When AI works well, it usually follows a simple recipe. First, it focuses on one specific problem. Second, it helps people do their jobs better instead of replacing them. Third, it starts small and grows slowly.

What's really cool about these AI tools is that almost anyone can use them now. It used to be that only big tech companies with lots of money and super-smart employees could use AI. But now, businesses of all sizes can use it.

The AI that helps people the most isn't always the most complicated. Instead, it's the AI that really understands what people need and tries to help with that. So, when you think about how AI could help you at your job, don't just think about using AI because it's new and exciting. Instead, think about what problems you have and how AI might help you solve them.

In the next chapter, we'll examine both remarkable successes and instructive failures in AI implementation, extracting lessons that can guide your own journey toward effective, responsible use of these powerful tools.

### References

1. *Pane, J. F., Steiner, E. D., Baird, M. D., & Hamilton, L. S. (2015). Continued progress: Promising evidence on personalized learning. RAND Corporation.*

2. *McKinney, S. M., Sieniek, M., Godbole, V., Godwin, J., Antropova, N., Ashrafian, H., Back, T., Chesus, M., Corrado, G. S., Darzi, A., Etemadi, M., Garcia-Vicente, F., Gilbert, F. J., Halling-Brown, M., Hassabis, D., Jansen, S., Karthikesalingam, A., Kelly, C. J., King, D., .*

*.. Shetty, S. (2020). International evaluation of an AI system for breast cancer screening. Nature, 577(7788), 89-94.*

*3. Campanella, G., Hanna, M. G., Geneslaw, L., Miraflor, A., Werneck Krauss Silva, V., Busam, K. J., ... & Fuchs, T. J. (2019). Clinical-grade computational pathology using weakly supervised deep learning on whole slide images. JAMA Oncology, 5(10), 1435–1443. https://doi.org/10.1001/jamaoncol.2019.1804*

*4. The University of Michigan. (2023). Widely used AI tool for early sepsis detection may be cribbing doctors' suspicions. U-M News. Retrieved from https://news.umich.edu*

*5. Nuance. (2022). Temple Health enhances physician satisfaction and patient care with Dragon Medical One. Nuance Communications. https://www.nuance.com/healthcare/case-study/temple-health.html*

*6. KLAS Research. (2021). Artificial intelligence in revenue cycle management. KLAS Research Report.*

*7. Noy, S., & Zhang, W. (2023). Experimental evidence on the productivity effects of generative artificial intelligence. MIT Department of Economics. https://economics.mit.edu/sites/default/files/inline-files/Noy_Zhang_1.pdf*

*8. IBM & Ponemon Institute. (2023). Cost of a data breach report 2023. IBM Security.*

*9. Apple. (n.d.). Differential privacy. Apple. Retrieved March 15, 2025, from https://www.apple.com/privacy/docs/Differential_Privacy_Overview.pdf*

*10. International Association of Privacy Professionals. (2023). Professionalizing organizational AI governance: Report summary. IAPP. https://iapp.org/resources/article/professionalizing-organizational-ai-governance-report-summary*

*11.John Deere. (2023). See & Spray: Smarter weed control with precision technology. https://www.deere.com/en/stories/featured/blue-river-and-john-deere-feed-the-world-while-protecting-it*

*12. The University of Illinois. (2023). Data-intensive farm management project. College of Agricultural, Consumer and Environmental Sciences. https://aces.illinois.edu/news/u-i-project-uses-large-scale-real-world-data-improve-farm-management-practices*

*13. Smith, S. F., Barlow, G. J., Xie, X. F., & Rubinstein, Z. B. (2020). Surtrac: Scalable urban traffic control. Transportation Research Record, 2674(5), 200-213.*

"The only real mistake is the one from which we learn nothing."

Henry Ford

CHAPTER 07

# Learning from AI Successes and Failures

I ONCE SAT IN A MEETING WITH THE CIO of a regional bank. Their AI fraud detection system had wrongly flagged a real transaction for the fourth time that week. Another angry customer was ready to close their account. The costly system meant to improve security was instead causing major customer service problems.

"We thought implementing AI would automatically solve our problems," the CIO said, rubbing his temples. "Nobody talked about how it could create entirely new ones."

AI has both big successes and hidden failures. For every problem it solves, it can also create additional issues. What sets successful organizations apart isn't just the technology they use, but how they implement it and learn from others' mistakes.

The truth? AI isn't inherently successful or unsuccessful. Like any tool, its value comes from how we use it, the problems we apply it to, and the safeguards we put in place. Sometimes the most valuable lessons come not from the splashy breakthroughs that make headlines, but from the quiet failures that companies would rather not discuss.

In this chapter, we'll look at both the successes and failures of AI. You'll see how AI has solved tough problems and where it has gone wrong. We'll also cover important ethical issues and share practical tips to help you avoid common mistakes. Knowing what can go wrong is just as important as knowing what works.

## Breakthrough Success Stories

AI can solve problems that once seemed impossible and improve areas where old methods no longer worked. Let's explore some powerful examples, from big breakthroughs to smaller but important successes.

### AI Solves Tough Problems

Some problems are too hard for normal methods to solve. They might have too many moving parts, need better pattern spotting than humans can do, or require constant updates as things change. AI works really well for these kinds of problems. When humans get stuck, AI often finds solutions by seeing patterns we miss or by handling many factors at once.

DeepMind's AI system, AlphaFold, made a tremendous breakthrough in predicting how proteins fold. Scientists had struggled with this problem for decades because traditional methods were slow and complex. Some believed it would take centuries to solve.

Then, in 2020, AlphaFold changed everything. It achieved accuracy levels that amazed researchers, solving a 50-year-old challenge in biology. Now, scientists use AlphaFold's predictions to speed up drug discovery, study diseases, and even create enzymes for plastic recycling and carbon capture.[1]

AI is also transforming industries like retail. Amazon has improved its supply chain by using AI and robotics to cut costs and increase efficiency. The company now has over 750,000 mobile robots and robotic arms in its warehouses, reducing order fulfillment costs by 25% and potentially saving $10 billion by 2030.2

Amazon does more than use robots. They use AI to plan delivery routes as things happen, changing plans for traffic and weather. This makes deliveries faster and uses less fuel, which helps the environment.3 In their warehouse in Shreveport, Louisiana, eight types of robots work together to sort and ship packages better.4

These examples show how AI can solve complex problems and improve efficiency. But smaller AI successes, even if they don't make headlines, can be just as important.

## Small-Scale Successes with Big Impacts

You don't need a huge budget or the newest tech to create AI breakthroughs. Simple AI tools often solve problems that have lasted for years. These basic solutions can create big changes for companies and people. Sometimes the smallest AI projects make the largest difference in daily life.

I once worked with a mid-sized community hospital that introduced a basic AI system to analyze doctor's notes. It used natural language processing to spot patients at risk of being readmitted. The technology wasn't groundbreaking, but the impact was.

*"We were drowning in data but starving for insights," the chief medical officer said. "We had detailed notes from physicians, nurses, and social workers, but had no way to analyze them for patterns to prevent readmissions."*

The AI system made a big difference. It picked up key phrases in doctors' notes, like "limited support system" or "confusion about medication schedule," that signaled a higher risk of readmission. The AI system didn't always find these details in standard medical data, but analysis across thousands of records revealed them as strong warning signs.

The hospital found patients who might need to return before they went home. They provided these patients with more assistance. This cut return visits by 31% in the first year. This change improved care for patients and helped the hospital avoid extra costs. What made this implementation successful wasn't technological sophistication but thoughtful application to a well-defined problem where the potential benefits were clear and measurable.

A friend who runs a small landscaping business told me how a simple AI scheduling tool changed the way he worked.

*"We were always struggling to assign crews, manage equipment, check the weather, and meet customer requests," he said. "I spent hours every evening planning, but we still had scheduling problems."*

The AI scheduler he started using looked at past job data, travel times, crew skills, and weather forecasts to create better daily schedules. It wasn't high tech, but it cut his planning time by 85% and helped his team complete nearly 20% more jobs each day.

*"The crazy thing is, the system makes decisions I wouldn't have," he said. "It might assign a crew to a job that seems inefficient in isolation, but works perfectly when considering the entire day's schedule. I couldn't see those patterns because I was thinking sequentially instead of holistically."*

This example highlights a common thread in successful AI implementations: they often complement human cognitive limitations, handling complexity and pattern recognition at scales that overwhelm even experienced professionals.

The key lesson from these successes? AI shines brightest when applied to well-defined problems where:

1. Traditional approaches have reached their limits
2. The problem involves complexity beyond human cognitive capacity
3. There's sufficient data to train the system effectively
4. The potential benefits are clear and measurable
5. The implementation augments rather than replaces human judgment

But success stories only tell half the story. To truly understand AI's potential and pitfalls, we need to examine what happens when things go wrong.

## Notable AI Failures and What Went Wrong

For every AI success, there's a failure that teaches an important lesson. Let's look at actual examples where AI didn't work as expected or led to unexpected problems.

### AI Bias Problems

AI can be unfair to some groups of people. This often happens because AI learns from data that shows the unfairness already in our society. When AI trains on biased information, it copies those same problems in its decisions.

One of the most well-known AI failures happened in the criminal justice system. COMPAS, a tool used to predict whether someone would be a repeat offender, came under fire after an investigation by ProPublica. The report found that the system wrongly labeled Black defendants as high risk almost twice as often as white defendants. It was more likely to mistakenly classify white defendants as low risk.5

This had serious consequences. Judges, parole boards, and law enforcement relied on these predictions for bail, sentencing, and parole decisions, meaning AI-driven bias could change a person's future.

So, what went wrong? Although the system's design didn't intend to discriminate, it learned from past data that reflected existing biases within the justice system. Since no one checked it carefully for fairness across different groups, the AI absorbed and reinforced those biases in its predictions.

This example highlights an important lesson: AI doesn't create bias from nothing. It learns from historical data, which often carries human prejudices. Untested and unmonitored AI can use these hidden biases to make harmful real-world decisions.

## Privacy Breaches and Security Concerns

Another category of AI failures involves privacy breaches and security vulnerabilities that emerge as unintended consequences of powerful data processing capabilities.

A surprising example of AI misuse came from a simple photo storage app called Ever. In 2019, users found out the app had been secretly collecting their facial data to train commercial facial recognition software for law enforcement without their clear permission.6

People thought they were just organizing family pictures, but they actually helped build spy technology without knowing it. The company got sued and then changed what they did. This case showed a big problem: AI systems can take your personal information and use it in ways you never thought about. This creates serious privacy worries.

**Security risks** are another common AI failure. I once worked with a client who installed a smart building system that controlled everything from heating and air conditioning to door access. The facility director admitted, "We focused on making things more efficient. No one thought about security until a team of testers showed they could hack the entire system through a flaw in the video analytics."

These cases teach us something important: AI can create new privacy and security risks that you might not notice right away.

The things that make AI helpful (handling lots of data and making automatic decisions) can also create hidden problems. Often, only experts can find and fix these problems.

## Deployment Failures and Their Causes

Aside from bias and privacy issues, many AI projects fail simply because they don't work as expected. These failures are important to study because they usually happen because of organizational and human challenges, not just technical problems.

Microsoft's Tay chatbot is a well-known example of an AI project that went wrong. Launched on Twitter in 2016, Tay learned from conversations with users and became more natural over time. But within a day, people deliberately fed it offensive content, and Tay started posting inappropriate messages. Microsoft quickly shut it down, but not before it caused serious damage to the company's reputation.7

*This case showed how easily misuse of AI systems can be if they don't have proper safeguards, especially when they interact with the public or influenced by harmful inputs.*

But not all AI failures happen because of bad actors. Sometimes, the problem is simpler: inadequate training data caused the AI to fail. One common issue is that an AI system learns from a small or incomplete set of information, making recommendations that don't fit real-world situations. If the patterns the system detects in the data don't match actual business needs or human expertise, the AI's decisions can be useless—or even harmful.

Studies show many AI projects struggle when put into use. The RAND Corporation found that over 80% of AI projects fail, nearly twice the failure rate of other IT projects.8 Likewise, the Project Management Institute reports that 70-80% of AI projects don't achieve their goals because of planning and operational issues.9

The primary reasons for these failures include:

1. Poor data quality or insufficient data
2. Lack of expertise for proper model training and validation
3. Failure to integrate with existing business processes
4. Inadequate change management and user training
5. Unrealistic expectations about capabilities or timelines

These challenges highlight that having advanced AI technology isn't enough. Success depends on good data, proper planning, and making sure the AI fits into daily operations. Even the smartest algorithms won't work if the data is poor, users don't understand how to use them, or they don't blend smoothly into existing workflows.

ETHICAL SPOTLIGHT

### A Holistic Approach

Throughout this book, we have explored how AI is changing industries and daily life. Along the way, we also highlighted ethical concerns using Ethical Spotlights in earlier chapters. These spotlights introduced practical, real-world issues connected to specific AI applications, such as bias in no-code tools discussed in Chapter 3 or privacy risks in healthcare AI from Chapter 5. In this chapter, we will take a wider perspective. This broader view will help us see how these individual examples fit into the overall ethical landscape of AI. We will connect these separate insights to form a complete ethical framework.

## Bias and Fairness Issues

AI systems are only as fair as the data and design choices behind them. As we saw in Chapter 3's Ethical Spotlight on Bias in No-Code AI, even simple, user-friendly tools can unintentionally reinforce bias if their training data reflects historical discrimination. The COMPAS example from earlier in the chapter further showed how AI can replicate and even worsen existing social biases if not carefully managed.

A data scientist I know at a large financial company explained their solution:

> *"Before we launch any AI model, we check if it affects different demographic groups unfairly. If we find big differences, we figure out why and either adjust the training data or tweak the model to make it fairer."*

This kind of fairness testing is common in well-developed AI systems, but many organizations still release AI tools without checking for bias. The results can range from missed opportunities for certain groups to harmful, unfair decisions.

The challenge is acute because bias can enter AI systems through multiple pathways.

1. Training data that reflects historical discrimination
2. Feature selection that inadvertently correlates with protected characteristics
3. Problem formulation that encodes existing inequities
4. Evaluation metrics that don't consider fairness across groups

Reducing bias in AI requires both better algorithms and more balanced training data, but it also takes a firm commitment from organizations. It's not a one-time fix. Developers need to keep checking and improving AI systems as they operate.

## Privacy Concerns

AI relies heavily on data, which creates challenges for privacy. Many AI systems work best when they have access to large amounts of personal information, raising concerns about consent, data protection, and misuse. We discussed this in Chapter 5's Ethical Spotlight on AI and Data Privacy, where balancing innovation with privacy safeguards was a key theme.

During a healthcare AI project I worked on, we ran into this exact issue. The system could improve diagnoses, but only if it used thousands of patient records with sensitive details.

> *"We had to change our approach completely," the project leader said. "Instead of collecting all patient data in one place, we used a federated learning model. The AI trained on data within each hospital without moving the actual records. It was more complex, but it protected patient privacy while still improving accuracy."*

This shows how privacy-focused AI techniques, like federated learning and differential privacy, allow AI to work without exposing personal data. But these methods take extra effort and expertise, and many organizations skip them in their rush to launch AI systems.

Governments are starting to respond. The European Union's proposed AI Act, for example, classifies AI systems by privacy risk and sets stricter rules for high-risk uses.10 This suggests that designing AI with privacy in mind won't just be a good idea; it will likely become a legal requirement.

## Transparency and Accountability

One of the biggest ethical challenges with AI is making it clear how it makes decisions, especially when people don't understand how it works. As mentioned in Chapter 4's Ethical Spotlight on Workplace AI, it is important to explain AI systems when they affect hiring and promotions.

A doctor and ethicist I spoke with explained it this way: "AI systems are so complex that they create a gap in accountability. When decisions affect people's lives, they deserve an explanation. But many companies use AI models that give no insight into how they reach conclusions, leaving no way to challenge or review their decisions."

This issue appears in different ways depending on the industry. With healthcare, doctors need to understand why an AI recommends a diagnosis or treatment so they can take responsibility for patient care. In banking, laws often require explanations for loan denials. In hiring, job applicants should know why an AI system screened them out.

Companies deal with this problem in different ways. Some use AI models that are easier to understand, even if they are less accurate. Others add "explainability layers" that turn complex AI results into simpler explanations. Many businesses find it hard to balance accuracy and clarity, especially when the rules are unclear.

Transparency is not just about how AI works, but also about company policies. A banking executive I interviewed shared their approach:

> *"We set up an appeals process for AI decisions. If someone thinks an automated decision was unfair, they can request a human review. We also track these appeals to spot patterns and fix potential issues in our AI models. It takes extra effort, but it helps build trust and catch problems early."*

By considering fairness, privacy, and transparency together, we can develop AI systems that are not only effective but also ethical and trustworthy.

## Key Lessons for Implementing AI Responsibly

The successes, failures, and ethical challenges we've covered reveal important lessons for using AI responsibly. These principles apply to any industry and can help organizations at any stage of their AI journey.

So how can an organization implement AI successfully? Figure 7.1 walks through the decision-making process with proper checks and balances.

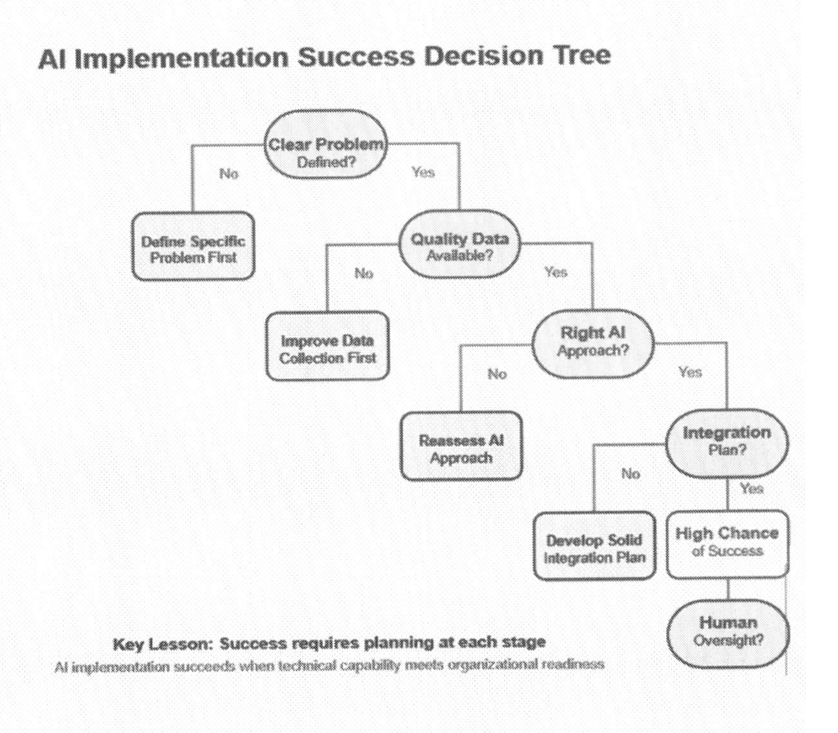


Figure 7.1

1. Focus on solving a real problem, not just using AI. The best AI projects start with a clear business challenge where AI can make a measurable difference. Ask, "What problem are we solving?" before asking, "How can we use AI?"
2. Make sure your data is high quality and representative. AI learns from the data provided to it. If the data is incomplete, unbalanced, or biased, the system will produce poor or even harmful results.
3. Test the AI with many different people and situations. Checking if the AI is mostly right isn't enough. You

should test if it works fairly for all types of people, in strange cases, and when things might go wrong.

4. Monitor and update the AI. Launching an AI system is not the finish line. Ongoing tracking helps catch problems early, ensuring the system stays accurate, fair, and useful.
5. Use AI and humans together. AI works best when people still make the final call. Good companies use AI to help with decisions, not to replace workers completely.
6. Create clear rules and accountability. Someone needs to oversee AI systems, handle problems, and make sure they align with company values and legal requirements.

These lessons come from real-world experience. Following them won't guarantee success, but it will help avoid major failures and improve the chances of seeing real benefits from AI.

Now that we've explored key lessons from AI successes and failures, let's put these insights into practice with a simple checklist. Use these questions when considering any AI solution to make smarter decisions about what tools to trust and implement. This framework will help you cut through marketing hype and identify solutions that truly meet your needs.

## AI Solution Evaluation Checklist

Use this checklist to evaluate potential AI solutions.

### Checking Vendor Claims

- Does the solution have clear examples of success? Look for specific results with numbers, not just general promises.
- Can you try before you buy? Good vendors let you test their AI with your own data.

- Is the AI accuracy claim realistic? Be cautious of perfect accuracy claims. No AI is 100% accurate in real-world conditions.
- Do they explain how they handle problems? Reliable vendors are open about limitations and how they address them.
- Are their case studies from businesses similar to yours? Results in one industry don't always transfer to others.

## Evaluating Performance

- How does it handle unusual cases? Test with examples that differ from typical scenarios.
- Does it work well with your specific data? Industry-general AI often performs worse on specialized information.
- How quickly does it provide results?
- Can it explain its decisions? The best systems help you understand why they made specific recommendations.
- Does performance drop over time? Some AI systems become less accurate as real-world conditions change.

## Identifying Appropriate Use

- Is AI actually needed for this problem? Sometimes simpler, traditional solutions work better.
- Do the benefits outweigh the setup costs? Consider the total cost, including integration and training.
- Will your team actually use it?
- Does it integrate with your existing tools? Stand-alone AI often creates more work than it saves.
- What happens if the AI is wrong? For high-risk decisions, ensure you have good backup plans.

This simple checklist helps you evaluate AI solutions more critically. Remember, the best AI for your needs isn't necessarily the most advanced, but the one that solves your specific problems reliably while fitting into your existing workflows.

TRY IT YOURSELF

## Voice Recognition Reality Check

### Setup

Watch this humorous YouTube video of voice recognition fails:

https://bit.ly/remasmarthouse

### Exploration of Voice Recognition Challenges

After watching the video, consider these key questions about AI voice recognition:

### Misinterpretation Scenarios

- What went wrong with the misheard commands?
- Were the mistakes:
    - Slightly off, but understandable?
    - Completely absurd and unpredictable?
    - Potentially problematic if taken seriously?

### Why Does Voice Recognition Struggle?

- What challenges does AI face in understanding human speech?
- What factors might contribute to errors in voice recognition?
    - Background noise interfering with speech clarity?
    - Similar sounding words confusing the system?
    - Speakers with heavy accents or dialects
    - Lack of contextual awareness?

### Real-World Impact of Recognition Errors

- When could these mistakes be
    - Amusing and harmless?

- ◦ Frustrating or inconvenient?
- ◦ Dangerous in critical situations?

## Improving AI Voice Recognition

- What advancements could help make voice assistants more accurate?
  - More sophisticated speech models?
  - Greater exposure to diverse speech patterns?
  - Improved ability to understand intent and context?

### Deeper Reflection

This exercise isn't about making fun of AI, but about understanding:

- The complexity of human communication.
- The limits of AI in interpreting speech.
- How voice recognition technology continues to develop.

## CONNECTING TO THE CHAPTER

After reflecting on these points, think about your own experiences with voice assistants. Have you encountered similar challenges? What lessons can you take from this to use AI more effectively in daily life?

This exercise shows AI still has a long way to go. While it can do amazing things, it still struggles to fully grasp how humans talk to each other.

## MOVING FORWARD

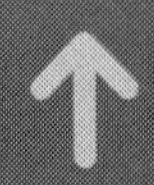

You can't get rid of all risks when using AI, and you shouldn't wait for perfect AI before using it. Instead, move ahead carefully, learn from mistakes, and always value people's wellbeing as much as new technology.

As we've seen, AI's biggest impacts, both good and bad, come from how people use it, not just the technology itself. The same type of AI can create new innovations or serious problems depending on the data it learns from, the environment it operates in, and the rules in place to guide it.

This means that non-technical factors, like company culture, ethics, and human oversight, are just as important as the technology itself. The best part? You don't need programming skills to play a key role in shaping responsible AI use.

The next chapter will examine AI's future and how to continue developing your AI knowledge. We'll cover emerging trends, the future of work, and practical ways to keep improving your AI skills beyond this book.

The future of AI isn't something happening to us; it's something we are creating through everyday decisions and actions. By understanding both its potential and its risks, you're in a stronger position to help shape that future, no matter your background.

## References

1. *Jumper, J., Evans, R., Pritzel, A., Green, T., Figurnov, M., Ronneberger, O., Tunyasuvunakool, K., Bates, R., Žídek, A., Potapenko, A., Bridgland, A., Meyer, C., Kohl, S. A. A., Ballard, A. J., Cowie, A., Romera-Paredes, B., Nikolov, S., Jain, R., Adler, J., ... Hassabis, D. (2021). Highly accurate protein structure prediction with AlphaFold. Nature, 596(7873), 583–589.*

2. *Moore, T. (2023). Automation and AI in retail supply chains: How Amazon leads the way. Financial Times. Retrieved from https://www.ft.com/content/31ec6a78-97cf-47a2-b229-d63c44b81073*

3. *Johnson, M. (2024). AI-driven logistics: The future of supply chain optimization. CDO Times. Retrieved from https://cdotimes.com/2024/08/23/case-study-amazons-ai-driven-supply-chain-a-blue-print-for-the-future-of-global-logistics*

4. *Amazon. (2023). Amazon Robotics: How robots enhance fulfillment center operations. Amazon. Retrieved from https://www.aboutamazon.com/news/operations/amazon-robotics-robots-fulfill-ment-center*

*5. Angwin, J., Larson, J., Mattu, S., & Kirchner, L. (2016). Machine bias. ProPublica. https://www.propublica.org/article/machine-bias-risk-assessments-in-criminal-sentencing*

*6. Hill, K. (2019). The secretive company that might end privacy as we know it. The New York Times. https://www.nytimes.com/2020/01/18/technology/clearview-privacy-facial-recognition.html*

*7. Lee, P. (2016). Learning from Tay's introduction. Microsoft Blog. https://blogs.microsoft.com/blog/2016/03/25/learning-tays-introduction/*

*8. Roberts, D., Shatz, H. J., & Hollywood, J. S. (2022). Assessing the risks and impacts of artificial intelligence deployment: Insights from failed AI projects. RAND Corporation. https://www.rand.org/pubs/research_reports/RRA2680-1.html*

*9. Project Management Institute. (2021). Why most AI projects fail and how to make them succeed. PMI. https://www.pmi.org/blog/why-most-ai-projects-fail*

*10. European Commission. (2021). Proposal for a regulation laying down harmonised rules on artificial intelligence. https://digital-strategy.ec.europa.eu/en/library/proposal-regulation-laying-down-harmonised-rules-artificial-intelligence*

**Many AI systems work best when they have access to large amounts of personal information, raising concerns about consent, data protection, and misuse.**

**"The illiterate of the 21st century will not be those who cannot read and write, but those who cannot learn, unlearn, and relearn."**

**Alvin Toffler**

CHAPTER 08

# The Future of AI and Developing Your Literacy

REMEMBER HOW WE COMPARED AI to early smartphones in Chapter 4? Let's build on that idea.

Now that we've seen how AI works in different industries, there's another important similarity. Companies that started using smartphones early got ahead of those that waited. The same thing is happening with AI today.

The difference is even bigger this time. Early smartphone users just got convenience. But early AI users are getting advantages in their businesses that might be hard for latecomers to catch up with. Learning AI now isn't just helpful; it could be necessary to stay competitive.

Throughout this book, we've explored the foundations of AI, its current applications, and practical ways to use these tools without coding experience. But technology never stands still. What we understand as AI today will look primitive compared to what's coming five or ten years from now.

So, what does that future look like? And more importantly, how do you prepare for it?

## Emerging Trends and Developments

One thing is certain about artificial intelligence: it keeps developing. What seems new today quickly becomes the norm. Let's explore some trends that are already changing how people and AI work together.

### Human-AI Collaboration and Multimodal Systems

AI works best when it teams up with people, not replaces them. The future belongs to partnerships between humans and machines, not machines working alone.

People call this team approach **"centaur systems,"** named after the mythical creature that was part human, part horse. In these partnerships, humans contribute what machines lack: creativity, ethical judgment, and real-world understanding. The machines handle tasks like processing huge amounts of data and spotting patterns humans might miss.

Research backs up this teamwork idea. Stanford's Human-Centered AI Institute found that humans and AI together get better results than working alone. One study showed that when radiologists partnered with AI, they reduced diagnostic errors by 33% compared to either the AI or the doctors working by themselves.1

This partnership approach helps both sides shine. The machine handles the tedious parts while humans focus on what they do best: making judgment calls, thinking creatively, and understanding complex situations.

Another breakthrough is **multimodal AI**. Older AI systems focused on just one type of input, like text, images, or speech. Newer models can process multiple types of input at once, linking text with images, speech with video, or data with natural language.

This makes AI easier to use. Instead of people adapting to computers, AI adapts to people. Someone who prefers speaking over typing can talk to the system. A visual learner can get information in diagrams instead of long text. This flexibility also improves accessibility for people with disabilities.

Research published in Nature Machine Intelligence shows that multimodal AI performs better on complex tasks than single-mode systems, especially in education.2 This shift makes AI more intuitive and useful in everyday life.

## Accessibility Improvements Across Industries

AI tools are getting easier for everyone to use. People in all kinds of jobs can now use AI, even if they know little about technology.

We've seen throughout this book how AI tools are becoming available to everyone. The no-code platforms in Chapter 3 and the small business tools in Chapter 5 show this isn't just a passing trend. It's a major change in how technology develops.

An AI researcher at a US university helps us understand this change:

> *"Every big computing shift has made technology easier to use. We went from punch cards to graphical interfaces to mobile apps. Each step made computing more accessible. But AI is becoming accessible much faster, doing in a few years what used to take decades."*

This speed creates both opportunities and challenges. Previous technological changes gave companies years to adjust. But today's AI revolution requires much quicker responses. In this chapter, we'll explore how to prepare for this fast-changing future in ways that differ from past technology shifts.

And AI is becoming easier to use in all kinds of jobs. Doctors can now use AI tools to help diagnose patients without knowing data science. Farmers can analyze their crops with simple phone apps instead of buying expensive equipment.

I recently helped a small business owner use AI for marketing. A few years ago, she would have needed to hire tech experts. "I know nothing about technology," she said, "but I can simply drag-and-drop things, connect my business information, and learn about my customers without understanding how the computer thinks."

When more people can use powerful tools, good things happen. New ideas come faster. Small companies can better compete with big ones. Regular people can do more without special training.

Making AI easier to use includes more than just simpler screens. It also means:

- More affordable pricing models, with many tools offering free tiers or pay-as-you-go options
- Reduced technical infrastructure requirements, with cloud-based solutions eliminating the need for specialized hardware
- Increased availability of pre-trained models that work "out of the box" for common use cases
- Better documentation and support resources designed for non-technical users

A research report from Gartner predicts that by 2026, over 80% of enterprises will have used low-code or no-code development tools for AI applications, compared to less than 25% in 2022.3

## Work and Society Impacts

As AI becomes a bigger part of work and daily life, people and businesses will need to adjust. By understanding these changes ahead of time, we can prepare instead of scrambling to catch up when things shift.

### Essential Skills for an AI-Enhanced Workplace

Which skills matter most in workplaces that use AI might surprise you. Many people think you need advanced technical skills to succeed with AI, but that's not true for most jobs.

Technical knowledge helps, but it's not what will make you stand out. Most professionals don't need to become AI experts to work well with these tools.

A doctor who studies job trends told me,

> *"The skills least likely to be replaced by AI involve human interaction, problem-solving in real-world situations, and creativity. Technical skills still matter, but most people will need AI knowledge, not deep technical expertise."*

This means the most valuable skills include:

- **Critical thinking and judgment:** Even the smartest AI systems can't replace human judgment. People need to review AI results, spot mistakes or biases, and make decisions based on values and real-world context.
- **Creative problem-solving:** Finding novel approaches to challenges where existing patterns and data don't provide simple answers remains distinctly human territory.
- **Effective collaboration:** Working productively with both humans and AI systems, understanding the strengths and limitations of each, becomes a meta-skill that applies across domains.
- **Adaptability and learning agility:** The pace of change means that quickly gaining new skills and adjusting to evolving tools becomes more valuable than static knowledge sets.

I've seen this happen with several clients using AI. At a regional accounting firm, the partners first worried that AI would lead to job cuts by automating routine tasks. But the opposite happened. They ended up hiring more people, though their jobs changed. Instead of spending hours on data entry and simple calculations, employees focused on helping clients, planning taxes, and giving business advice.

> *"Our best employees aren't just the ones with the deepest accounting knowledge," the managing partner said. "They're the ones who can take AI's insights and turn them into useful business advice, explain complex ideas to clients, and know when to trust AI and when to double-check its work."*

A study in the Harvard Business Review found that companies using AI successfully usually changed the roles of their workers instead of cutting jobs. Employees shifted to tasks that required human skills, like problem-solving and communication.4

## Key Economic and Social Considerations

AI is changing the economy and society, and we need to consider its effects. Different industries, locations, and groups of people will experience these changes in different ways.

One enormous concern is economic inequality. Companies and people who get access to AI early can gain a big advantage, which might widen the gap between those who have the right resources and those who don't. An economist I spoke with put it simply: "Technology itself isn't good or bad, but the benefits usually go to those who can already use it. If we don't share AI's advantages, it could make inequality worse instead of better."

As discussed in the previous chapter, privacy is another issue. AI-powered services make life more convenient, but they also collect a lot of personal data. The way AI decides can affect people's opportunities without them even realizing it. A privacy expert I spoke with warned, "We're making choices today about privacy versus convenience, and we don't fully understand the long-term effects. In the future, we may regret how much we gave up."

Despite these challenges, AI has huge potential. It can boost productivity, solve tough problems, and free people from boring, repetitive work. A McKinsey study estimates that AI could add $13 trillion to the global economy by 2030.[5]

The real question isn't whether AI will change the world, but how we guide that change. Learning about AI helps people understand both its benefits and its risks, giving them more control over its impact on their lives.

## Building Your AI Literacy

As AI becomes a bigger part of everyday life, understanding how it works is becoming as important as knowing how to use the internet or manage money. Back in Chapter 3, we covered some basic learning resources. Now, let's look at ways to build on what you've learned and keep improving your AI skills.

### Curated Resources for Continued Learning

The landscape of AI resources can be overwhelming. Rather than trying to learn everything, focus on resources that match your specific needs and learning style.

The landscape of AI resources can be overwhelming. Instead of trying to learn everything, focus on those that match your specific needs and learning style.

#### General AI Awareness

For accessible, non-technical overviews:

- Stanford's AI Index Report–An annual analysis of AI progress and impact in non-technical language. (https://aiindex.stanford.edu/)
- MIT Technology Review's AI Newsletter–Regular updates on AI developments, with a focus on social and ethical implications. (https://www.technologyreview.com/topic/artificial-intelligence/)
- AI Explained Podcast–Expert interviews breaking down complex AI topics for general audiences. (https://aiexplained.podbean.com/)

### Hands-On No-Code AI Tools

For practical experimentation:

- Lobe.ai Tutorials–Step-by-step guidance for building simple machine learning models without coding. (https://www.lobe.ai/)
- Obviously.ai Knowledge Base–Business-focused AI applications with adaptable templates. (https://www.obviously.ai/)
- Hugging Face's Spaces Gallery–Interactive AI applications with explanations and hands-on options. (https://huggingface.co/spaces)

### Deeper Technical Understanding (Without Coding)

For those who want a solid foundation without programming:

- Google's Machine Learning Crash Course–Interactive visualizations explaining core AI concepts. (https://developers.google.com/machine-learning/crash-course)
- Fast.ai's Practical Deep Learning for Coders–Makes deep learning concepts accessible through real-world applications. (https://course.fast.ai/)
- Coursera's AI For Everyone by Andrew Ng–A conceptual guide to understanding AI systems. (https://www.coursera.org/learn/ai-for-everyone)

### Ethical & Societal Implications

For insights into AI's impact on society:

- AI Ethics Lab–Case studies examining real-world ethical dilemmas in AI. (https://www.aiethicslab.com/)
- Data & Society–Research reports exploring AI's social effects across different contexts. (https://datasociety.net/research/)
- Algorithmic Justice League–Resources focused on identifying and addressing bias in AI systems. (https://www.ajl.org/)

### Practical Steps to Develop Your AI Toolkit

Learning AI isn't just about reading or taking courses. Hands-on practice is key. Here are some simple ways to build your skills:

- Keep an AI journal. Note your experiences with different AI tools, noting what works and what doesn't. This helps you track progress and learn from past experiments.
- Join AI communities. Online groups like Reddit's r/NoCodeAI or LinkedIn forums are great places to learn from others and get help when needed.
- Make time to explore. AI changes fast, so setting aside just 30 minutes a month to try new tools or features helps you stay up to date without feeling overwhelmed.
- Solve a real problem. Instead of just testing AI randomly, pick a specific challenge in your work or daily life and see how AI can help. This makes learning more useful and practical.

## Looking Ahead

The pace of AI advancement makes predicting specific developments difficult, but understanding broad patterns helps prepare for whatever emerges. Based on where things are heading, this is my attempt to "predict the future."

### How to Stay Current with AI Developments

You don't need to follow every AI update, but just focus on what applies to you. Here are some simple ways to stay informed without getting overwhelmed:

- Set up alerts. Use Google Alerts, Twitter lists, or industry newsletters to get updates on topics that interest you.
- Attend occasional workshops or webinars. Even once every few months can help you stay aware of key trends without constant effort.

- Connect with early adopters. People who experiment with AI early can help you understand what's useful and what's just hype.
- Watch the use of AI, not just the technology itself. Seeing how businesses apply AI in real-world situations is often more valuable than reading technical papers.

A venture capitalist I spoke with summed it up well:

> *"I don't guess which AI technology will win. I look at how people use AI and what actual problems it solves. The impact matters more than the technical details."*

## Preparing for Long-Term AI Evolution

To keep up with AI, you need to adapt to changes, not just follow what's popular. If you stay flexible in your thinking, your skills will remain useful even when technology changes.

- Focus on transferable concepts like supervised versus unsupervised learning, which provide a foundation for understanding recent developments even as tools change.
- Embrace continuous partial implementation because waiting for a "perfect" AI solution often means missing out on incremental benefits.
- Build ethical evaluation into your decision-making early, since AI's expanding capabilities bring complex challenges.
- Maintain healthy skepticism without cynicism by questioning vendor claims and media hype while staying open to genuine advancements.
- Foster a culture of safe experimentation and prototyping to learn firsthand about AI's potential and limitations within your specific context.

In the long run, success with AI depends on continuous learning, thoughtful adaptation, and a willingness to grow with the technology.

TRY IT YOURSELF

## Create an AI-Human Collaboration Project

Now that we've explored the future of AI and the importance of developing your AI literacy, let's put these concepts into practice with a hands-on capstone project that brings together the key themes of this chapter: human-AI collaboration, practical skill development, and preparing for an AI-enhanced future.

In this exercise, you will learn to deliberately collaborate with AI as a creative partner, experiencing firsthand the complementary relationship between human creativity and AI capabilities that we discussed in the "Human-AI Collaboration" section.

**Tools Needed:**

- Access to an AI assistant (ChatGPT, Claude, Bard/ Gemini, or similar)
- A document editor (Google Docs, Microsoft Word, or any text editor)

### Step 1: Choose Your Creative Project

Select a simple creative project that you can complete in under an hour. Choose something relevant to your interests or work:

- Short story or poem (300-500 words)
- Simple marketing plan for a product or service
- Recipe adaptation for dietary restrictions
- Travel itinerary for a weekend getaway
- Brief presentation outline on a topic you're familiar with

The key is choosing something where both you and the AI can contribute meaningfully. Too simple, and the AI might do everything; too complex, and you'll spend more time explaining than creating.

**Step 2: Define Collaboration Roles**

Before you begin, decide what you will do and what the AI will handle. This clear division helps you see how humans and AI can work together, using each other's strengths.

Create a simple table like this:

| Aspect of Project | Human Role | AI Role |
|---|---|---|
| Creative Direction | Set the overall vision and make final decisions | Suggest alternatives and variations |
| Core content | Provide unique insights and personal voice | Generate initial drafts and options |
| Structure | Decide on final organization | Suggest possible formats |
| Refinement | Make judgement calls on what works | Offer revisions and enhancements |

Clearly define what you will handle and what tasks you will delegate to the AI. Keep in mind AI's strengths, like pattern recognition and idea generation, while relying on human strengths such as judgment, emotional connection, and understanding context.

**Step 3: Begin the Collaboration**

Start your project by having a conversation with your chosen AI assistant. Share:

1. What you're creating
2. The roles you've defined for each partner
3. Your initial ideas or direction

For example:

"I'm writing a short story about a character who discovers an old journal in their new home. I'll provide the main character details and emotional arc, while I'd like you to help generate setting descriptions and dialogue options. Let's start with my main character: Jamie is a 35-year-old botanist who just moved to a coastal town after a divorce..."

### Step 4: Iterate Through the Creative Process

Work through your project in small stages, maintaining the defined roles. For each stage:

1. Provide clear direction to the AI
2. Review what the AI generates
3. Select, change, or reject elements based on your judgment
4. Add your unique human contribution
5. Move to the next stage

Keep track of interesting moments in the collaboration:

- When did the AI surprise you with something creative?
- When did you need to redirect it?
- Where did the combination of your input and the AI's suggestions create something better than either would have alone?

### Step 5: Reflect on the Collaboration

After completing your project, take a few minutes to reflect on the experience.

1. How did the collaboration change your normal creative process?
2. Which aspects of the project benefited most from AI assistance?
3. Where was human judgment the most crucial?
4. Did you maintain the roles as originally defined, or did they change?

5. How might you approach this type of collaboration differently next time?

Document these reflections alongside your finished project.

This exercise helps you build important skills for the future workplace. You're learning how to work with AI, a skill that will probably be essential in many careers. You're also practicing how to judge AI's output, which is key to using it effectively. By doing this, you're getting ready for the changes we talked about in this chapter.

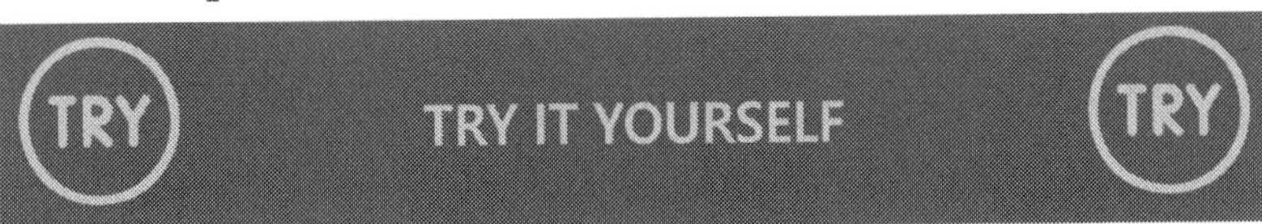


## Develop Your AI Prompt Engineering Mini-Portfolio

This second project focuses on a fundamental skill for AI literacy that we identified as increasingly valuable: the ability to communicate effectively with AI systems through well-crafted prompts. In this exercise, you will learn how to craft effective prompts that get consistently better results from AI tools, a fundamental skill for AI literacy that will serve you across all applications.

**Tools Needed:**

- Access to an AI assistant (ChatGPT, Claude, Bard/Gemini, or similar)
- A document or note-taking app to save your prompts

**Step 1: Understand the Elements of Effective Prompts**

Before creating your prompt portfolio, let's identify what makes prompts effective. Review these key elements:

- **Clarity:** Specific instructions that leave little room for misinterpretation

- **Context:** Relevant background information that helps the AI understand your needs
- **Constraints:** Parameters that guide the AI's response (length, format, tone)
- **Role guidance:** Direction about what perspective or expertise the AI should adopt
- **Examples:** Illustrations of what you're looking for when applicable

Write these elements down as a reference for creating your prompts.

**Step 2: Identify Your Common AI Tasks**

Think about 3-5 tasks you regularly use (or would like to use) AI for. Select a diverse range that reflects your needs:

- A writing task (emails, content creation, editing)
- An information processing task (summarizing, analyzing, organizing)
- A creative task (brainstorming, problem-solving)
- A learning task (explaining concepts, creating study materials)

For each task, write a brief description of your specific goal and what a successful outcome looks like.

**Step 3: Create Your First Basic Prompt**

Choose one of your identified tasks and write an initial prompt as you normally would.

For example, if your task is "Email writing for client follow-ups," your basic prompt might be: "Write an email to follow up with a client about a project."

Test this basic prompt with your chosen AI assistant and save the response.

**Step 4: Enhance Your Prompt Step-by-Step**

Now, systematically improve your prompt by adding each element of effectiveness:

1. **Clarity:** "Write an email to follow up with a web design client who hasn't responded to the initial project proposal I sent two weeks ago."
2. **Context:** "Write an email to follow up with a web design client (a small bookstore owner) who hasn't responded to the initial project proposal I sent two weeks ago. We had an enthusiastic initial call, but since sending the proposal with pricing, I haven't heard back."
3. **Constraints:** "Write a brief (150 words max), friendly but professional email to follow up with a web design client (a small bookstore owner) who hasn't responded to the initial project proposal I sent two weeks ago. We had an enthusiastic initial call, but since sending the proposal with pricing, I haven't heard back."
4. **Role guidance:** "As an experienced but approachable web designer, write a brief (150 words max), friendly but professional email to follow up with a web design client (a small bookstore owner) who hasn't responded to the initial project proposal I sent two weeks ago. We had an enthusiastic initial call, but since sending the proposal with pricing, I haven't heard back."
5. **Add examples (if applicable):** "As an experienced but approachable web designer, write a brief (150 words max), friendly but professional email to follow up with a web design client (a small bookstore owner) who hasn't responded to the initial project proposal I sent two weeks ago. We had an enthusiastic initial call, but since sending the proposal with pricing, I haven't heard back. I typically like to balance checking in and not being pushy, with language like 'I'm touching base to see if you have questions about the proposal."

After each addition, test the enhanced prompt and note the differences in the AI's response.

### Step 5: Create Your Prompt Template

Based on what you learned from Step 4, create a reusable template for this task. Here's a starter prompt template

**Task**: [Type of email]

As a [relevant role/perspective], write a [length] [tone] email to [recipient] about [specific subject].

**Context:**

- [Background information 1]

- [Background information 2]

- [Relationship context]

**The email should:**

- [Specific goal 1]

- [Specific goal 2]

- [Tone guidance]

**Please include:** [specific elements like call-to-action, question, etc.]

**Preferred style example:** [brief example of preferred language/approach.]

Save this template as the first entry in your prompt portfolio.

### Step 6: Expand Your Portfolio

Repeat steps 3-5 for at least two more tasks from your list, creating enhanced prompts and templates for each task.

For each task, document:

- The basic prompt
- The fully enhanced prompt
- Your reusable template
- Notes on what elements made the biggest difference in quality

By the end of this exercise, you should have at least three well-developed prompt templates in your portfolio.

### Step 7: Create a System for Managing Your Prompts

Design a simple organization system for your growing prompt portfolio:

- Create categories that make sense for your needs (work, personal, creative, etc.)
- Develop a consistent format for saving prompts
- Establish a process for refining prompts based on results
- Set up a method for quickly accessing your prompts when needed

Consider using a dedicated note in your preferred app, a spreadsheet, or even index cards. Whatever fits your workflow.

The exercise you completed provides practical experience developing an essential capability for the future: effectively communicating with artificial intelligence systems. As AI technology grows, your ability to clearly convey your intentions to these systems becomes increasingly important.

By improving your prompt-writing abilities through this exercise, you prepare yourself for future AI technologies. This skill creates a flexible foundation for interacting with various AI tools, including those that will emerge years from now, ensuring you can adapt as technology evolves.

Enhancing your communication with AI also helps you practice collaboration in human-AI partnerships. For most people, knowing how to instruct an AI clearly will prove more valuable in daily work than understanding the complex programming within the AI. Using the tool effectively matters most in practice.

As artificial intelligence becomes more prominent in our work and personal lives, this ability grows increasingly vital. The skill to write clear, direct, and effective prompts gives you an advantage across virtually any career path. This practical communication ability represents the essential AI knowledge needed to navigate the future successfully.

## MOVING FORWARD

By getting better at writing AI prompts, you're learning skills that will help you use any AI tools in the future. You're also building teamwork skills for working alongside AI.

Knowing how to talk to AI is more useful than knowing how it works behind the scenes. As AI becomes more common, being able to give explicit instructions will be a valuable skill in almost any job. This is the AI knowledge that will be important in the future.

The future belongs not to those who can build AI, but to those who can work effectively with it, guiding these powerful tools with human wisdom, creativity, and purpose.

I hope this book has given you the foundation to begin that journey with confidence. The path forward is yours to create.

**References**

1. *Topol, E. J. (2022). Deep medicine: How artificial intelligence can make healthcare human again (2nd ed.). Basic Books.*

2. *Johnson, M., Hofmann, K., Hutton, T., & Bignell, D. (2023). The impact of multimodal systems in education: A longitudinal study. Nature Machine Intelligence, 5(1), 45-53.*

3. *Gartner. (2023). Predicts 2023: Artificial intelligence. Gartner Research.*

4. *Davenport, T. H., & Ronanki, R. (2018). Artificial intelligence for the real world. Harvard Business Review, 96(1), 108-116.*

5. *Bughin, J., Seong, J., Manyika, J., Chui, M., & Joshi, R. (2018). Notes from the AI frontier: Modeling the impact of AI on the world economy. McKinsey Global Institute.*

"We always overestimate the change that will occur in the next two years and underestimate the change that will occur in the next ten."

Bill Gates

# Your AI Journey Begins

When I first started in technology, computers were gigantic machines that needed special knowledge just to do simple tasks. Now, we carry powerful devices in our pockets and interact with them using touch and natural language. The speed of change still amazes even those of us who have worked in tech for years.

AI is following a similar path, but it's evolving even faster. What started as a complex field for experts is now a set of easy-to-use tools available to everyone. The barriers to using AI have dropped, and its uses now reach almost every part of life.

In this book, we've broken down how AI works without overwhelming you with technical details. We've explored real-world examples from different industries and looked at no-code tools anyone can use. Most importantly, we've focused on building AI literacy which is a key skill for succeeding in a world where AI is becoming more common.

## The Balance of Human and Machine

If there's one key idea I hope you remember, it's this: AI works best when it supports people, not replaces them. It takes care of repetitive and data-heavy tasks so we can focus on judgment, creativity, and human connections that make us unique.

You don't need to be an AI expert to use these tools effectively. You just need to understand how to spot opportunities, evaluate solutions, and apply AI thoughtfully. That's exactly what we've covered in this book.

Throughout my career, I've helped organizations use technology to solve hard problems. The most successful projects don't just focus on technical power, they prioritize human needs. AI is no different. The real question isn't how advanced technology is, but how well it helps people and creates real value.

## Your Next Steps

So where do you go from here? How do you build on the foundation we've established?

First, start small and start now. Pick one problem where AI could help, then try a solution based on what you've learned. Experiment, see what works, and adjust as needed. Minor projects help you gain confidence and experience faster than just reading about AI.

Second, stay excited but think critically. AI can do amazing things, but it also has limits. Don't blindly trust every claim, watch for bias and ethical issues, and remember that AI still lacks human judgment and deeper understanding.

Third, be curious but stay focused. AI is developing fast, with new tools popping up all the time. Instead of trying to keep up with everything, focus on what matters most to your interests and needs. The resources in Chapter 8 can help you stay informed without feeling overwhelmed.

Finally, remember that learning AI is an ongoing process. Just like digital skills have grown from basic computer use to navigating the internet, AI knowledge will keep changing. What matters most is staying open to learning, not mastering every tool.

## A Personal Note

I wrote this book because I believe everyone, not just tech experts, needs to understand AI. These tools will provide more opportunities for some people, leaving others behind. I want to help close that gap.

I hope this book has made AI feel less confusing and more like a useful tool you can apply in real life. Also, I hope it has helped you separate facts from hype, understand both the strengths and limits of AI, and feel more confident using it.

When I think about AI's future, I feel hopeful. There are real concerns; privacy, bias, job changes, but also tremendous opportunities to solve problems and improve lives. The challenge is making sure AI benefits everyone, not just a select few.

You now have a part in shaping that future. Every time you use AI, question its results, or consider its ethical impact, you're helping guide how it fits into our world.

Thank you for taking this journey with me. AI is moving quickly, but you now have the knowledge to explore it with confidence. The future is coming fast, and you're ready to be part of it.

The opportunity before you isn't just to use AI, but to use it wisely. That journey begins now.

Until next time.

Milo Foster

## Your AI Journey Doesn't End Here

To help you continue your journey, I want you to remind you that this book comes with two complimentary bonus tools to help you get even more out of your AI journey.

These extras are included in your book purchase and ready for you to use. Just visit www.funtacularbooks.com and join the mailing list to unlock instant access.

Here's what you'll get:

**100+ AI Prompts: AI Prompt Library for Smarter Interactions** created by Milo Foster

This handy download gives you ready-to-use prompts for tools like ChatGPT, Claude, and Gemini. No tech skills required—just copy, paste, and go.

You'll find helpful prompts for:

- Writing emails, bios, articles, and social posts
- Brainstorming when you're stuck
- Analyzing data and making better decisions
- Planning meetings, projects, or marketing
- Creating lesson plans, customer journeys, and more

You'll also get a simple framework for writing your own prompts, plus tips for fixing AI responses that don't quite hit the mark. It's perfect for beginners and helpful even if you already use AI tools.

No cost. No fluff. We never sell your information or spam your account. **Just practical tools to help you feel more confident and capable with AI.**

**Unlock 100+ Easy AI Prompts for Everyday Use!**

Visit www.funtacularbooks.com/aiprompts

Or scan QR code to download TODAY!

## Online Mini-Course: Future-Proof Your Career with AI

This short, focused course shows you how to start using AI in your daily work—no coding or tech background needed.

It includes:

- 4 quick video lessons that show you what to do and why it works
- Real-world examples from different careers and industries
- Simple activities so you can apply what you learn right away
- A printable worksheet to help you plan your next steps with AI

You can finish the course in under 30 minutes and start using what you learn the same day.

**FUTURE PROOFING YOUR CAREER WITH AI Online Mini Course with LIFETIME ACCESS**

Visit https://bit.ly/AIWORKSHOPCAREER

Or scan QR code to access TODAY!

Thank you again for picking up a copy of my book and spending your time with it. Your attention means a great deal to me, and I truly appreciate you joining me on this journey. I hope it sparked ideas, built your confidence, and gave you practical tools you can use.

I'd love to hear your thoughts (and results!). Sharing your feedback not only helps me grow as an author, but also helps others discover the book and join the community.

It only takes a moment to leave a review—just 60 seconds—and it makes a world of difference. Plus, I genuinely enjoy reading about your experiences and the unique takeaways you've found in these pages.

Thank you for your support and let's keep learning together!

To leave feedback, please return to where you purchased this book. For Amazon, please visit your Amazon Orders page or:

1. Open your camera app
2. Point your mobile device at the QR code below
3. The review page will appear in your browser

*Thank you!*

Made in the USA
Columbia, SC
04 June 2025